Ghosts of the Hills

Ghosts of the Hills

Jack Mercer Chronicles

Todd Robertson

To my wife,

my partner, my inspiration, and my greatest supporter.

Thank you for everything.

Table of Contents

Prologue

Korea, January 1952

The wind howled across the ridgeline, a banshee's wail that cut through wool and flesh alike. Snow whipped sideways, blinding, erasing the world in white and shadow. Staff Sergeant Jack Mercer crouched low behind a rock shelf, breath freezing in his scarf, the frost riming his lashes. His Thompson rested across his knees, the receiver crusted with ice.

Below, through the curtain of snow, the valley pulsed with faint orange light—Chinese lanterns and cook fires flickering behind earthen berms and tangled wire. A forward command post, small but alive. The kind of place that moved nightly, too valuable to hit head-on, too slippery to leave alone.

Mercer raised his field glasses, the glass fogging instantly. He grunted, wiped them on his sleeve, and scanned again.

"Two sentries, north approach. At least a platoon dug in around those huts," he murmured.

Sergeant Davis, huddled beside him, muttered through chattering teeth, "Whole damn army of 'em for a few crates of rice."

Corporal Hill, the oldest of the bunch and perpetually sour, spat into the snow. "Or maybe something worth more than rice. Intel says the Reds've been moving radio parts through this sector. Could be what command wants."

"Command always *wants* something," Reilly said, the team's Irish demolitions man, his grin faint under a frostbitten face. "Just a question of how many of us they're willing to spend to get it."

Kim, the quiet scout, pointed toward a dark gap between the ridges. "Trail runs behind the position. If we cut over the saddle, we can get eyes on whatever they're guarding without getting chewed up."

Mercer nodded once. "That's the play. We move in five."

No one argued. They'd been together for too long for that—too many nights crawling through mud, snow, and blood. They weren't the clean-cut recon boys the brass liked to parade in briefing photos. They were the misfits—what was left after the disciplined ones cracked or froze or didn't come back. Mercer's crew didn't polish boots or salute much anymore. They just got the job done.

As the men rose, Hill clapped Mercer on the shoulder. "You ever get tired of this, Sarge? Ten years of cold fronts and bad odds?"

Mercer's eyes flicked toward the valley below, the lines of light and movement dancing like ghosts in the storm.

"Yeah," he said quietly. "But tired don't mean done."

They slipped into the dark shadows among shadows, moving single file through the snow and pine. The world narrowed to breath, heartbeat, and the slow crunch of boots. Somewhere far off, artillery thudded like distant thunder—another part of the line bleeding.

Mercer adjusted his scarf, feeling the weight of years heavier than his pack. The uniform had changed, the enemy's tongue was different, but the war always looked the same.

And as the night swallowed them whole, he knew one thing with cold certainty.

War might have moved on, but Jack Mercer never did.

Part 1: Echoes of War

Chapter 1: The Long Road Home

It had been five years since the guns fell silent. Peace had come to the world, but not to Jack Mercer. He walked to the edges of the Millersville farm, feeling the wind stir the tall Iowa grass, and felt the strange weight of his own life pressing against him. Anytime he started to grow close to someone, he reminded himself: he would outlive them by decades. One day, inevitably, he would have to move on.

Since 1945, Jack had drifted from place to place, job to job, searching for something he could never name. He had returned to Millersville, hoping for the comfort of family and the familiar fields of his youth. But the farm wasn't the same. His brother had died at Leyte Gulf, and the empty space he left behind haunted every corner of the homestead. Jack knew he could not bear to watch his parents' age, their lives moving forward while his body remained untouched by the years.

The war had left scars no one could see. Worse, the lab in Normandy had left him with secrets darker than any battlefield. The serum. The alteration. One year aged for every ten lived. His reflection betrayed none of it, yet inside, decades of memory, loss, and survival weighed on him like armor he could not shed.

Jack tried civilian life, tried to anchor himself. He worked as a ranch hand in Kansas, mending fences and breaking horses. He drove trucks across the Southwest, sleeping in his cab while towns flashed by in a blur. For a time, he became a bounty hunter, tracking fugitives across deserts and small towns, tasting a fraction of the tension and stakes he had once known in combat. Each attempt at normalcy failed. Restlessness was constant, gnawing in his chest, whispering that he was not meant for this slow, ordinary life.

Even in Millersville, in the quiet of the early morning, memories of war haunted him. He remembered Sicily, 1943: heat, dust, and the roar of artillery. Sam, cautious and watchful, flanked him; Doc adjusted his rifle; Rizzo, reckless, scanned the rooftops.

"Move," Sam had whispered. "Keep it tight."

The ambush had come suddenly and brutally. Men fell around them. Jack survived, guided by training and the strange endurance that made him different. Later, in a captured German bunker, the serum had been injected. One year aged for every ten lived. Time had become a strange, distant thing — and Jack knew he would never be like other men again.

Now, standing alone on the familiar land, he watched the horizon. The world kept moving, but he was stuck outside of it, a soldier without a war, a man out of step with time itself. He had survived battle after battle, yet surviving peace proved to be something else entirely.

Jack Mercer lit a cigarette, letting the smoke curl into the morning air. Somewhere beyond the horizon, the next chapter of his life waited — restless, unknown, and inevitable.

The morning sun beat down on the Central Texas spread, glinting off the dry grass and the worn metal of the fences Jack had been repairing. He moved efficiently, mending broken posts and tightening wires, muscles memorized to the rhythm of labor. Work always helped him think—or, more accurately, kept him from thinking too much.

A rider appeared on the dirt road, kicking up a cloud of dust. Jack straightened, shading his eyes, and recognized the urgency in the man's face before he spoke.

"Jack?" the rider called. He handed over a telegraph.

Jack tore it open and read the words, the typewriter script sharp and final:

Jack—Your Mom and Dad were in a car accident. Not expected to survive. Come home immediately. Cousin Lisa.

Time seemed to slow. Jack's chest tightened, though outwardly he remained calm. He folded the wire and tucked it into his pocket, staring out across the endless fields. Millersville. The farm. His parents. He had avoided thinking about them growing old, but now that avoidance had been ripped away.

He didn't speak to the rider. There was no need. Jack mounted his horse, the animal sensing the tension in its rider. His mind raced with practicalities: packing, travel arrangements, timing—but underneath, a flood of emotions he rarely allowed himself to feel dread, guilt, and helplessness.

The ranch, the fences, the cattle all faded behind him as he rode hard, thinking of the last time he had been home. He remembered the farmhouse, the smell of fresh-cut hay, and the quiet evenings at the kitchen table with his parents and his brother before the war.

Now it is gone, or about to be. And Jack Mercer, whose body barely aged and whose mind had survived horrors no one else could imagine, felt the weight of mortality in a way that no war or experiment had ever forced upon him.

Jack tied the reins of his horse to the hitching post, brushing sweat and dust from his forearms. The ranch house loomed in the morning haze, the scent of horses, hay, and sunbaked dirt heavy in the air. He glanced up at the wide porch where Mr. Handley, the ranch owner, waited.

"Mr. Handley," Jack said, approaching. His voice was calm, even though inside, a storm brewed. "I just got word from Millersville… my parents were in an accident. They're not expected to make it. I need to go home."

Mr. Handley's face tightened, a mixture of concern and understanding passing across it. He nodded slowly, wiping his hands on a rag. "I'm sorry to hear that, Jack. I know Millersville's been your home… or was, anyway. You take care of your family first."

Jack hesitated, feeling a pang of guiLT "I don't want to leave the ranch in a bind. The fences, the cattle…"

Handley waved a hand dismissively. "Don't worry about it. I'll make sure things keep running. Here." He pulled a small stack of bills from his pocket and handed them to Jack. "This will cover your travel and anything else you need. Take care of yourself, and… take care of them."

Jack pocketed the money, his fingers brushing the worn leather of his beLT Relief mixed with the familiar tug of restlessness. Even now, duty and ties to others felt like chains he had learned to manage cautiously.

Handley stepped closer, placing a firm hand on Jack's shoulder. "You know, Jack… you always have a place here on the Bar H Ranch. No matter where life takes you, this place will always be yours."

Jack nodded, a faint, almost imperceptible smile crossing his face. "I appreciate that, sir. I really do."

He mounted his horse once more, dust rising around the hooves. The Bar H Ranch had given him stability, a place to rest, but he knew in his bones he couldn't stay for long. There was a world moving on without him, and Millersville called.

With one last look at the sprawling ranch, Jack urged his horse forward. The road ahead was long, and the past, with all its burdens and secrets, waited for him.

Jack led his horse to the small tack shed, the familiar scent of leather and wood filling the air. He removed the saddle and bridle carefully, hanging each piece on its peg, then rubbed the horse down with a coarse brush, feeling the heat and sweat of the animal under his hands. The horse nuzzled him gently, sensing the tension in its rider.

"You've been a good partner, buddy," Jack murmured, running a hand along the animal's neck one last time. The routine, the rhythm of work he had done countless times before, grounded him for a moment. But even as he finished, a knot of urgency tightened in his chest.

He stepped into the bunkhouse, where the cool, shadowed interior smelled of dust, leather, and worn blankets. Jack opened his trunk and began packing: a few changes of clothes, his personal gear, a worn travel bag. Every movement was methodical, almost ritualistic, as if preparing for the unknown journey ahead could somehow give him control over the world he could not predict.

Jim Handley, the oldest son of Mr. Handley, lounged in the corner of the bunkhouse, arms crossed, watching Jack with a mixture of curiosity and respect. "You'll need a ride?" he asked casually, though the concern in his eyes was clear.

Jack nodded, zipping the travel bag closed. "Yeah. Train station. I need to get moving as soon as possible."

Jim stood and stretched, moving toward the door. "Ranch truck's warmed up. I'll get you there." He hesitated a moment, then added, "You're always welcome back, Jack. Don't forget that."

Jack gave a short nod, appreciation flickering across his features but tempered by the restlessness that never left him. "I know, Jim. I'll remember."

With the horse settled and his belongings packed, Jack slung the travel bag over his shoulder and stepped out of the bunkhouse. The early morning light cut across the ranch, dust motes floating in the beams, a final, quiet farewell before the road pulled him away toward Millersville, and the family waiting there.

The ranch truck rumbled along the dirt road, dust curling behind the tires. Jack sat in the passenger seat, travel bag on his lap, eyes fixed on the passing plains. Jim Handley drove with quiet efficiency, hands steady on the wheel, scanning the horizon. For miles, neither spoke, letting the low hum of the engine fill the silence.

Jack thought of the last five years drifting from job to job, town to town, trying to find a place to belong and never quite managing it. Ranch work, truck driving, bounty hunting, odd jobs in cities he barely remembered—all temporary, all fleeting. Home had been a word he avoided, a place that reminded him of loss, of his parents' inevitable aging, of the brother he had lost at Leyte Gulf.

Jim glanced at him. "You alright?"

Jack nodded, jaw tight. "Yeah… just thinking." He turned his gaze back to the horizon. "Life moves fast, even when you're standing still."

Jim gave a short nod, understanding without needing words. The road stretched on, empty except for the truck and the wind through the prairie grass. Jack felt the familiar tug of restlessness, tempered now by urgency—the fragile thread of family pulling him home.

After a long drive, the landscape shifted from wide-open plains to the small buildings of Marble Falls. Jim pulled the truck up to the train station, the familiar wooden platform and its weathered roof coming into view. The station was quiet this morning, the whistle of distant locomotives echoing faintly in the distance.

Jack exhaled, shouldering his travel bag. The journey ahead would no longer allow him to drift. The train would carry him toward Millersville, toward his parents, toward the fragile reality of life he usually skirted. He stepped out of the truck, feeling the weight of years—war, experiment, restlessness—pressing down, and followed Jim toward the platform.

Jack sat on the wooden bench of the Marble Falls train station, the early morning sun baking the platform, dust clinging to his boots. His travel bag rested at his feet as the faint whistle of a distant train echoed across the valley, mingling with the hum of cicadas.

As he waited, his mind drifted back to 1943, to a time before the war had fully claimed him. He had brought his squad—Sam, Doc, Rizzio, and Shorty—home for a brief leave before shipping out. The Mercer farmhouse had smelled of fresh hay and cooked breakfast, sunlight streaming through the windows. His parents had welcomed his friends as if they were family—his father laughing with Sam, Doc joking with his mother, Rizzio and Shorty fitting seamlessly into the rhythm of home.

The memory tightened something in Jack's chest. Those bonds had survived war and distance, and now he faced a new urgency.

He pulled a pen and a telegraph form from his bag, his hand steady despite the tension in his chest. He wrote carefully:

Doc, Sam—Just wanted to let you know my parents were in an automobile accident. They are not expected to survive. Thought you should know. Jack.

He handed the form to the station clerk, who nodded and took it to the telegraph office. Jack folded his hands, staring down the tracks that stretched toward the horizon. Sending this wire felt like a tether to the past, a way to reach out to the men who had once been brothers in arms and had always belonged, in a sense, to his family.

The train whistle sounded closer now, signaling the car that would carry him back to Millersville—and the uncertain reality waiting there. For a moment, Jack let himself dwell on memory, on family, and on friends who had been like brothers.

The sun was high over Sam's tobacco farm In Kentucky, casting long shadows between the neat rows of bright green leaves. Sam wiped sweat from his brow, leaning on a hoe as he surveyed the fields. Life on the farm was steady, predictable—the exact opposite of the chaos he had known with Jack and the squad during the war.

A rider appeared on the dirt lane, carrying a telegraph. Sam took it from the messenger, reading the crisp, typewritten words with growing concern:

Doc, Sam—Just wanted to let you know my parents were in an automobile accident. They are not expected to survive. Thought you should know. Jack.

Sam felt a tightness in his chest. The calm of the Kentucky fields, the familiar rhythm of planting and tending, could not shield him from the sharp pang of worry. Jack's parents had always been like a second family, and Sam knew the man himself—the restless soldier who rarely let anyone in—would be carrying a heavy weight on this trip back home.

He folded the wire carefully and pocketed it. He glanced toward the farmhouse, where the late summer breeze rattled the shutters. Thoughts raced trains to catch, arrangements to make, and the long drive from the farm to the nearest station.

Sam climbed into the old pickup, the engine rumbling to life beneath him. He checked the maps, planned his route, and knew he had to move quickly. Jack had always been the one to lead them into chaos, but now it was Sam's turn to answer the call, to be there for his friend and for the family he had known all those years ago.

As he rolled down the dusty lane, Sam felt the familiar mix of urgency and purpose. The farm faded behind him, the tobacco leaves swaying in the wind, and the journey toward Millersville began—toward reunion, toward comfort, and toward the fragile reality of mortality that even soldiers could not escape.

Doc sat at a small desk in the barracks at Fort Knox, the clatter of typewriters and distant shouts of drill echoing across the training grounds. The steady rhythm of military life was a comfort after years in combat—structured, predictable, and orderly. That comfort vanished the moment a telegraph arrived bearing Jack's familiar, terse handwriting.

He tore open the envelope and read the words quickly:

Doc, Sam—Just wanted to let you know my parents were in an automobile accident. They are not expected to survive. Thought you should know. Jack.

Doc's fingers tightened around the paper. The calm of Fort Knox, the routines of drills and schedules, seemed suddenly trivial. Jack's parents had always been like family to him too, and he knew the weight of this news would hit Jack hard.

He leaned back, exhaling slowly, and pulled out a notepad to make quick plans. The train schedules, travel routes, and arrangements to leave Fort Knox—all had to be organized immediately. There was no time to waste.

Doc rose from his desk, slinging a travel bag over his shoulder. The tanks and rifles around him reflected the world he would leave behind for a short time. He paused, thinking of Jack and the Mercer family, and knew he had to be there.

Every mile would bring him closer to Millersville, to the Mercer farmhouse, and to the uncertain reality waiting for Jack. Doc walked toward the gate, determination in his step. The journey had begun.

The whistle of the train pierced the dry morning air at the Marble Falls station, a long, resonant note that stirred something deep in Jack. Dust swirled around the platform as he hoisted his travel bag onto his shoulder and stepped toward the waiting car.

He paused for a moment, watching the train's steel frame glint in the sun. Boarding felt almost ceremonial, a threshold he had crossed countless times in the past—moving from one place to another, chasing work, chasing distance, chasing restlessness.

Jack settled into a window seat, placing his bag on the floor. Outside, the Texas plains stretched endlessly, the early morning light brushing the rolling hills and distant trees. His mind drifted to the past five years since the war had ended: ranch hand in Texas, truck driving across the Southwest, bounty hunting in small towns, fleeting stints in cities he barely remembered. Each place was a waypoint, a brief stop in a life that refused to settle, a life still searching for something he couldn't quite name.

The telegraph he had sent to Doc and Sam weighed lightly in his pocket—a thread reaching back to the men who had been his brothers in arms, the men who had shared in laughter, danger, and the rare moments of home during the war. The thought of seeing his parents again, even under grim circumstances, pulled him forward, anchoring him with a purpose he hadn't felt in years.

The train began to move slowly at first, then picking up speed. The clatter of wheels on rails blended with the rhythmic rocking of the car. Jack stared out the window, watching the landscape blur, feeling the familiar mixture of anticipation and unease. Home was waiting, but so was the confrontation with the fragile reality of life—his parents, his family, the small town that had always been both comfort and burden.

Jack leaned back in his seat, letting the train carry him forward. The past and present intertwined, memories of war and home mingling with the urgency of the present. The journey to Millersville had begun—and with it, a reckoning he could no longer avoid.

The small station outside Fort Knox In Elizabethtown buzzed with quiet activity as soldiers and civilians alike moved along the platform. Doc leaned against a post, travel bag slung over his shoulder, eyes scanning the tracks for the approaching train. The morning air carried the faint smell of coal and diesel, mingling with the scent of dew on the grass.

The whistle of the train pierced the quiet, drawing closer, and the steel cars rumbled to a slow halt along the platform. Doc adjusted his bag and stepped forward as the conductor signaled the doors.

As he climbed aboard, Doc's eyes caught a familiar figure settling into a seat further down the car. Sam.

"Sam?" Doc called softly, crossing the aisle.

Sam looked up, a wry smile breaking across his sun-weathered face. "Doc. Didn't think we'd meet on the same train," he said, nodding in greeting.

Doc grinned faintly, feeling the ease of old friendship settle around him like a familiar coat. "Guess fate's running the schedule today."

They shook hands firmly and took their seats together. Outside the window, the station faded as the train began to move, rolling past fields and small towns of Kentucky. The rhythmic clatter of wheels on rails seemed to echo the memories of war, of home, and of the many roads they had traveled since.

Both men fell into a companionable silence, each lost in thought. They knew Millersville awaited—a farmhouse, family, and the fragile reality of Jack's parents' condition. The journey stretched ahead, long and uncertain, but for the first time in years, Doc and Sam would face it together.

The train picked up speed, leaving Elizabethtown behind, carrying them eastward toward Millersville, toward Jack's family, and toward the reunion that none of them could yet imagine in full.

The train rattled steadily along the tracks, carrying Jack through the Texas plains and the rolling hills of Louisiana and Mississippi, and Doc and Sam through Kentucky, all converging toward the same destination.

Jack sat by the window, his travel bag at his feet, watching the landscape blur into streaks of green and gold. Memories of Millersville pressed in—the farmhouse, the smell of fresh-cut hay, the sound of his parents' laughter mingling with the voices of Sam, Doc, Rizzio, and Shorty years ago. He pressed a hand to the window, as if reaching for a past that now felt fragile and fleeting.

On another train, Doc and Sam exchanged quiet glances, each lost in thought but aware of the unspoken bond, tying them together. They had fought battles far from home, survived more than most could imagine, and now the war had ended— but life still demanded courage of a different kind.

The news of Jack's parents' accident weighed heavily on both of them, spurring a determination to be there for their friend.

The rhythmic clatter of the train was almost hypnotic, a steady heartbeat carrying them all toward Millersville. Each mile was a step closer, each passing landscape a reminder that no matter how far they had traveled or how restless the years had made them, some things—the family, the bonds of friendship—remained untouchable.

Jack leaned back in his seat, exhaling slowly. Ahead lay uncertainty, grief, and perhaps reconciliation. But for now, there was motion, there was purpose, and there was the shared, unspoken promise that they would face whatever awaited together.

The train whistle blew, sharp and clear, cutting through the afternoon air. The journey home had begun—and nothing would be the same again.

Chapter 2: The Empty Chair

The late afternoon sun cast long shadows over the small Millersville train station, the scent of warm dust and coal smoke hanging in the air. Sam and Doc stepped down from the station platform, stretching their legs after the long ride from Kentucky. The familiar hum of the town, quiet but steady, seemed almost unreal after the miles of open road and distant fields.

Doc adjusted his travel bag and glanced around. "We should find out when Jack's train is due," he said. His voice carried the mix of impatience and concern that came naturally to him.

Sam nodded, striding over to the station attendant behind the ticket counter. "Afternoon," he greeted. "Can you tell us what time the train from Texas will be arriving?"

The attendant shuffled through a stack of papers, eyes scanning the schedule. "Texas train? Let's see… right on time, it should be rolling in around 5:15. Give or take a few minutes."

Sam exchanged a glance with Doc. "He always lets us know where he's coming from," Sam said quietly. "We'll know he's on this one."

Doc leaned back against the counter, shoulders tense. "Good. Wouldn't miss it for the world." His eyes drifted to the tracks, where the rails gleamed in the sun. A long, quiet anticipation settled over them—both men remembering years of shared battles, laughter, and loss, and now the delicate gravity of what awaited their friend and his family.

The two of them stood together in companionable silence, watching the distant horizon. Somewhere out there, Jack was riding the rails from Texas, heading home, and soon they would all converge at the same place that had once been a sanctuary, a haven of family and friendship.

Every moment ticked by slowly, measured by the faint echo of the train whistle and the distant rhythm of approaching wheels. Millersville awaited—and with it, the reality of what Jack had left behind and what they were about to face together.

The distant whistle of the train carried across the fields, sharp and insistent, cutting through the calm of late afternoon. Jack pressed his forehead against the cool window of the car, watching the familiar landscape unfold—rolling hills, farmhouses, and the occasional barn he remembered from childhood. Each landmark tugged at memories he had tried to keep at arm's length over the years.

The train slowed, the rhythmic clatter of wheels on rails growing louder, mingling with the hiss of steam and the faint murmur of passengers preparing to disembark. Jack's travel bag rested at his feet, yet it felt heavier than usual, loaded with more than clothing—it carried worry, anticipation, and the weight of news he wished he didn't have to face.

He spotted the station platform, small and familiar, and saw the figures of Sam and Doc already waiting, leaning casually against the rails yet scanning every car that rolled into view. Relief and warmth rose in him, a rare grounding in the restless drift of his postwar life.

Jack's heartbeat faster as the train screeched to a haLT He grabbed his bag, swinging it over his shoulder, and stepped into the station, dust and heat rising around him. Sam waved, Doc gave a slight nod, and Jack returned a tight, grateful smile.

The three men converged quickly, greeting each other with brief, firm handshakes. Words weren't necessary, years of shared experience and loyalty spoke louder than anything.

Jack took a deep breath, glancing past them toward the town beyond the tracks, toward the familiar streets and the farmhouse where his parents lived. Every step felt heavy with urgency, yet the sight of his old friends by his side made the coming moments slightly more bearable.

The train had delivered him home. Now It was time to face what awaited in Millersville.

The three men lingered on the station platform, scanning the small town beyond the tracks, when a voice called out behind them.

"Jack! Sam! Doc!"

They turned to see a young woman stepping briskly toward them, a wide smile on her face. Lisa Mercer, Jack's cousin, now a young woman but still carrying the spark of the thirteen-year-old girl who had once run through the farmhouse kitchen, approached with open arms.

Without hesitation, she threw her arms around Jack, then Sam and Doc, hugging each of them warmly. "I've missed you all," she said softly, her voice tinged with both joy and relief.

Sam chuckled, ruffling her hair gently. "You've grown up, kid."

Doc nodded, smiling beneath his usual stoic demeanor. "Hard to believe you were just a girl when we were last here."

Lisa laughed lightly. "I still think of you both as family," she said, looking between Sam and Doc. "Jack might be my cousin, but you two… you've always been like brothers to me."

Jack watched the interaction with a mixture of warmth and unease. He could see how much his friends meant to Lisa and how much his aunt and uncle had loved these men all those years ago. The weight of the news about his aunt and uncle pressed at the edges of his mind, but seeing Lisa here—strong, vibrant, and unwavering—gave him a small, grounding comfort.

"You'll come with me," Lisa said, gesturing toward the town beyond the tracks. "Aunt and Uncle… they're at the hospital, and they've been asking about you."

Jack nodded, feeling the pull of family stronger than the restless drift that had marked the past five years. Together, the three warriors and the young woman who had grown up waiting for them started walking toward the hospital, toward the fragile reality that awaited.

The heavy double doors of the hospital creaked open as Jack, Lisa, Sam, and Doc stepped into the lobby. The sterile smell of antiseptic and the quiet hum of fluorescent lights replaced the warm dust and sunlit streets of Millersville. Nurses bustled past, pushing carts, answering phones, and guiding patients and visitors down hallways.

Lisa led the way, her pace brisk but steady. "They're in room 207," she said quietly, glancing at Jack. "I'll stay with you."

Sam and Doc came to a halt near the waiting room, exchanging a brief look.

"You go on ahead, Jack," Sam said. "We'll wait here for you. Don't need us crowding the room."

Doc nodded, a faint smile tugging at his lips. "Go on, brother. Take your time. Just focus on them. We'll be right outside if you need anything."

Jack hesitated for a moment, feeling the weight of anticipation pressing on his chest. Then he nodded, giving both men a tight, appreciative smile. "Thanks," he said softly.

Lisa guided him down the hallway, their footsteps echoing softly on the polished floor. Jack's mind raced with memories of home, of his mom and dad, and of the fragile thread between life and death that now tied him to this moment.

As they reached the door to room 207, Jack took a steady breath. He placed a hand on the handle, bracing himself for the sight of the people who had always been his anchor, now lying vulnerable and uncertain behind the thin hospital walls.

The waiting room faded behind him. For Jack, the next moments would be a reckoning—and a reminder that some battles were not fought with weapons, but with presence, courage, and heart.

Jack pushed open the door to room 207. The quiet hum of the hospital seemed to vanish, replaced by the sterile scent of antiseptic and the faint beeping of monitors. His mom lay in one bed, pale and frail, her hand resting limply on the blanket. His dad was in bed beside her, also pale, a bandage wrapped around his forehead, his chest rising and falling with effort.

Jack froze for a moment, his chest tightening. Both looked so fragile, unlike the strong, vibrant parents he remembered. Fear and sorrow churned in his stomach as memories of home and childhood collided with the stark reality before him.

"Jack?" His mom's voice was barely a whisper, thin and hoarse, yet filled with recognition.

Jack crossed the room in a few long strides, dropping to his knees between the two beds. "Mom… Dad…" His voice cracked as he reached for their hands. They were warm, but their weakness made his heart ache.

His dad's eyes lifted, straining to focus. "Jack… you're here," he said, voice faint, but the relief in his tone was unmistakable.

"I'm here," Jack replied, gripping their hands tightly. The fragile threads of their presence, the vulnerability in the room, made the reality of the accident hit him fully. "I'm not leaving."

Lisa stood just inside the doorway, silently watching, a comforting presence at his side. Sam and Doc lingered in the waiting room, giving Jack the space to face this moment alone.

Jack sank into a chair beside his mom's bed, brushing a damp strand of hair from her forehead. "I was so worried…" he murmured, voice trembling.

His dad tried to squeeze his hand, a faint but determined gesture. "Our boy… you're here… that's all that matters."

Jack's chest tightened again, a mix of relief and helplessness. Both his parents were so fragile, their strength diminished by the accident, and yet they still reached out to him, anchoring him in this room, in that moment.

He had come home. And now, he would stay—if only to be the strength they could no longer summon themselves.

Jack took a deep breath, keeping his hands gently clasped around his parents'. "Mom… Dad…" he began softly, "Sam and Doc—they're just outside. They came all the way to see you."

For a moment, his parents' weary eyes widened with recognition. A faint smile flickered across his mom's lips. "Sam… Doc…?" she whispered, her voice thin but filled with warmth.

His dad's hand, weak but steady, gave a small squeeze. "Our boys… we'd love to see them," he said, his voice barely above a whisper.

Jack looked at Lisa, who nodded, understanding immediately. "Lisa, can you get them for us?" he asked gently.

Lisa moved quickly, slipping out of the room. Within moments, Sam and Doc appeared in the doorway, pausing as Jack's parents' eyes met theirs.

Sam stepped forward first, a grin spreading across his weathered face. "Ma'am, sir… we're here," he said softly.

Doc followed, his usual stoic expression softened with emotion. "Even if it took us a while to get here," he added, giving Jack a brief nod before focusing on the parents.

Both Jack's mom and dad lifted themselves slightly in their beds, struggling but smiling, their eyes shining with gratitude and recognition. "Come closer," his mom murmured, reaching a trembling hand toward Sam.

His dad extended his hand toward Doc, and despite their frailty, the reunion carried the weight of years, memories, and battles shared. Jack stood a moment back, letting the warmth of the room—the laughter, the tears, and the unspoken bond—wash over him.

For the first time since the accident, Jack felt a fragile sense of hope. His parents were hurting, but they were still surrounded by the family and brothers-in-arms who had always mattered to him the most.

Sam and Doc stepped back slightly, letting Jack stay with his parents, their voices dropping to quiet murmurs as they spoke near the doorway.

"Remember those five days on the Mercer farm before we shipped out?" Sam said with a faint smile, shaking his head.

Doc chuckled softly. "How could I forget? Jack had us running sunup to sundown. Relentless PT—morning drills, long hikes, obstacle courses through the fields… and don't forget the hay bale runs," he added, a laugh in his voice.

Sam nodded, grinning. "He made us curse and sweat like madmen, but damn if it didn't pay off. I don't think I'd have survived the war if it weren't for those five days."

"Jack made us all the best versions of ourselves," Doc said quietly, his eyes scanning the room where Jack was holding his mom and dad's hands. "He pushed, he taught, he didn't let anyone slack. And somehow, he did it with that grin of his, like he enjoyed every second of it."

Sam laughed softly. "Yeah… we hated him at the time, but now? I'd follow that brother anywhere."

Doc nodded, resting his hand lightly on Sam's shoulder. "We all owe a lot to him. Those five days on the farm were brutal, but they made us ready. Made us strong. Made us brothers."

Sam glanced toward Jack, still sitting beside his parents. "And look at him now… still the same but somehow carrying all that with him. That boy's got fire, I'll tell you that."

Doc's smile softened. "That fire saved us more times than we can count. And seeing him here, with his folks… reminds me why it mattered."

Sam and Doc stood quietly for a moment, letting the memory settle between them, the weight of years and battles giving way to a sense of pride—and the unshakable bond forged back on that small farm, long before the world demanded they become soldiers.

Summer, 1943.

The train whistle echoed through Millersville's small depot. Jack stepped down first, uniform sharp and duffel slung over his shoulder. Sam, Doc, Shorty and Rizzo followed close behind, boots hitting the platform in unison.

Waiting at the end of the platform was Jack's mother, her apron still dusted with flour. She didn't hesitate, she wrapped her arms around Jack first, then pulled Sam and Doc into the same embrace as if they were her own. When she reached Rizzo, she gave him the same fierce hug, ignoring his mock-protest about "crushing a paratrooper's ribs."

Jack's father stepped forward next, offering a firm handshake to each man, his grip steady and warm. "Any friend of Jack's is family here," he said. "You'll have a bed, a hot meal, and a place at our table as long as you need it."

The four young men walked up the road together toward the Mercer farm, already feeling like they belonged.

—The memory faded, and the sound of gravel beneath the truck tires brought Jack back to the present. The porch swing still hung in its place, swaying faintly in the evening breeze, but there were no figures waiting there this time—no flour-dusted apron, no strong handshake. And no Rizzo, cracking wise about the smell of the barnyard.

Jack parked by the barn, his jaw tight. Sam and Doc climbed out beside him, their eyes lingering on the house in silence.

"Let's get our bags inside," Jack said finally, his voice steady but heavy. "Then we'll see what Lisa's stocked in the kitchen."

They walked toward the farmhouse, each step feeling heavier than the last, the ghosts of 1943 trailing close behind.

The night had settled over the farm in a thick, warm hush. Crickets chirped in the fields, and the faint rustle of leaves carried on the breeze. The porch boards creaked under the weight of three tired men, boots stretched out, shoulders slouched.

Jack handed each of them a tin cup filled with a splash of his father's old rye whiskey. The burn in his throat felt almost welcome.

They didn't speak at first. The silence wasn't uncomfortable, just heavy, like a well-worn blanket they all knew how to carry.

Sam stared out across the dark yard, the faint outline of the barn just visible in the moonlight. "Feels strange without him," he said finally.

Doc nodded, his thumb tracing the rim of his cup. "Rizzo would've had us laughing by now."

Jack glanced at the far end of the porch where an old rocking chair sat in shadow. The same spot Rizzo had claimed in '43, spinning tall tales about Brooklyn and swearing the country air made him stronger. The chair rocked slightly in the breeze, empty now.

They each took a slow sip, letting the memory sit between them. No toasts, no speeches, just three brothers-in-arms remembering the fourth.

Jack set his cup down on the porch rail. "We should head back," he said quietly.

Sam and Doc nodded. They stood, the boards creaking once more under their boots, and walked down the steps toward the truck. The night seemed to follow them as they drove off, the porch light glowing behind them like a small, stubborn beacon.

Chapter 3: Final Watch

The morning sun had just begun to burn away the thin veil of mist clinging to the fields as Jack, Sam, and Doc headed back toward Millersville. The old pickup's tires crunched along the gravel road, dust trailing behind them in the still air. None of them said much—the rumble of the engine was enough to fill the silence.

Jack sat in the passenger seat, eyes fixed ahead, but his thoughts were already at the hospital. He kept seeing his mother's-tired smile, his father's pale face. Sam drove with an easy hand on the wheel, but kept glancing at Jack, reading the tension in his jaw. Doc sat in the middle, leaning forward slightly as if he could make the truck go faster by will alone.

When they rolled into town, the streets were quiet—just a few shopkeepers setting out their displays, the scent of fresh bread drifting from the bakery. Sam parked outside the hospital, and the three men climbed out, their boots striking the wooden sidewalk in a steady rhythm.

Inside, the air was cool and smelled faintly of antiseptic. Jack led the way down the hallway toward his parents' room, the sound of their footsteps echoing against the tile. He didn't knock—just eased the door open, heart pounding as he stepped inside to be at their side once more.

Jack moved to his mother's side first, taking her hand gently, then turned to his father, giving a reassuring squeeze. Sam and Doc lingered near the doorway, silent but present, their eyes soft as they watched Jack with his parents.

"How's the farm?" his mother asked weakly, her voice barely above a whisper. "Everything still standing after all these years?"

Jack smiled faintly, though his chest tightened. "Yes, Ma. The barn, the house, even the porch swing is still there. It's… just like you remember it."

His father nodded, a tired but proud smile on his lips. "And the crops? The livestock?"

"Doing well," Jack replied. "I stopped by with Sam and Doc last night. We ate on the porch, watching the sun set over the fields. Felt like old times—except there's an empty chair," he said quietly, glancing at the corner where the memory of Rizzo lingered in his mind.

His mother's hand tightened around his. "You've done good, Jack. We're proud of you. And your friends… these boys who've been with you all these years—they're family too, aren't they?"

Jack nodded. "Yes, Ma. They've always been like brothers."

Sam stepped forward, his voice low and steady. "We've been with him through a lot… and we'll stay right here with him now."

Doc added, giving a small smile, "We're not going anywhere. Your boy's done more than anyone should ever have to, and we're proud to stand by him—and by you."

Jack's parents both smiled faintly, their eyes warm despite the frailty in their bodies. In that moment, the weight of years, wars, and loss felt a little lighter surrounded by family and the brothers who had shared every battlefield with their son.

Jack stepped back slightly, letting Sam and Doc continue their quiet conversation with his parents. He walked over to the window, gazing out at the distant fields beyond Millersville, imagining the Mercer farm stretching under the summer sun. The memory of those days—the laughter, drills, the sunburned afternoons—pressed gently against his chest.

He thought of Rizzo, always quick with a joke, always the first to challenge Jack during their makeshift drills on the farm. He should have been here with them, sharing in the quiet moment, swapping stories, teasing Sam or Doc. But he wasn't, and the empty space beside the porch rocking chair haunted him more now than ever.

Jack's gaze drifted back to his parents' pale faces. They were fragile, yes, but they were here, alive, and for the moment, at peace. He had wandered after the war, drifting from place to place, never settling, never staying. Yet somehow, all of his restless years had led him back to this—back to the people he loved, back to the brothers who had shared every battlefield, and back to the memory of the friend who had been taken too soon.

A small sigh escaped him, and he leaned against the window frame. The past and present coexisted in the quiet hospital room, the farm, the war, the friends lost and survived—and for the first time in a long while, Jack let himself simply be present.

He turned back to Sam and Doc, who had fallen silent as well, each lost in their own thoughts. Together, they were more than survivors—they were a bond that death, distance, and time could not break.

Jack smiled faintly, a shadow of the old fire in his eyes. He would carry the memories, the losses, and the love with him—and he would face whatever came next with them by his side.

Lisa glanced at Jack, Sam, and Doc as they gathered their coats. "I'll stay here with Aunt Jan and Uncle Bill," she said softly. "Make sure they're comfortable. You three go ahead and get some food—you've earned it."

Jack nodded, grateful. "Thanks, Lisa. Don't worry—we'll be back soon to check in."

Sam clapped her gently on the shoulder. "We'll try not to eat too fast, but no promises."

Doc chuckled, tipping an imaginary hat. "Take care of them, kid. They're in good hands with you."

Lisa smiled faintly, lingering at the doorway a moment longer before settling by her aunt and uncle's bedside. The three men stepped out into the afternoon sun, the warmth on their shoulders a small comfort after the tension of the day.

A few blocks away, the café awaited— familiar smells of frying bacon and fresh bread, and the promise of a proper meal. Jack, Doc, and Sam walked together in silence at first, each lost in their own thoughts, yet comforted by the quiet companionship of men who had been through every kind of battlefield together.

The bell above the café door jingled as Jack, Sam, and Doc stepped inside. The warm aroma of frying bacon, fresh coffee, and baked bread wrapped around them like a familiar hug.

Jack slid into a corner booth, Sam and Doc taking the seats across from him. The waitress, smiling at the three men, set menus down before quickly returning with a pot of coffee.

Jack took a slow sip of his steaming cup, letting the warmth and flavor settle over him. For the first time that day, he could feel the tension in his shoulders ease a little, grounding comfort after the weight of the hospital.

Doc stirred his coffee thoughtfully, his gaze distant. "You know… this right here," he said, glancing at the plates of hot food, "is something I didn't think I'd taste for a long time after the war. That POW camp… cold, thin meals, nothing warm or filling for days. You never appreciated a simple slice of bread until you went hungry enough to dream of it. And now… this," he nodded toward their plates, "this is a blessing."

Sam grinned, shaking his head. "Amen to that. A slab of bacon never felt so precious."

Jack smiled faintly, letting the warmth of the coffee and the camaraderie wrap around him. For a few minutes, the weight of worry over Aunt Jan and

Uncle Bill, and the memories of distant battlefields, eased.

They ate in companionable silence, occasionally exchanging a smile or a small comment about the food or the town. Even after years of war and loss, moments like this reminded them that life still held grounding pleasures worth saving.

The last of the coffee cups were emptied and the plates cleared away. Jack, Sam, and Doc rose from the booth, stretching stiff limbs and adjusting their coats. The comfort of the café lingered, but the pull of the hospital and Jack's parents awaited them.

Outside, the late afternoon sun had begun its slow descent, casting long shadows across the quiet streets of Millersville. The air was warm, carrying the faint scent of blooming flowers from nearby yards. Jack led the way to the truck, opening the door for Sam and Doc with a practiced ease.

Sam climbed in first, glancing over at Jack. "Nothing like a proper meal to remind you that some things are still worth fighting for," he said with a small grin.

Doc leaned back in the seat, his hands resting on the wheel. "Yeah… and makes you appreciate every little comfort. Makes the long nights and hard days of the war feel even more precious now."

Jack nodded, glancing at both men. "Let's get back. Mom and Dad will be expecting us, and Lisa's keeping them comfortable."

The truck rumbled to life, tires crunching over the gravel, carrying the three men back toward the hospital. Each was quiet, lost in thought, the comfort of the café still lingering, mingling with the weight of the day. For Jack, the drive was a reminder that even after years of war, loss, and uncertainty, there were still moments of peace—and moments of family—to hold onto.

As the hospital came into view, Jack's grip on the steering wheel tightened slightly, a mix of anticipation and concern passing over him. They were heading back to the bedside of those who mattered most, carrying with them both the warmth of home and the solemn reality that time was fleeting.

The three men stepped quietly into the hospital room. The soft hum of machinery and the faint smell of antiseptic greeted them, but the familiar sight of Jack's parents in their beds brought a sharper, more immediate focus.

Jack moved first, kneeling at his mother's bedside and taking her hand gently. "Mom… Dad…" His voice caught slightly, but he forced it steadily.

His mother's eyes brightened when she saw him. "Jack… you're here." Her weak smile carried all the warmth of home.

Jack's father stirred, lifting his hand slowly to rest on Jack's arm. "Good to see you, son. And… you brought friends," he added, glancing at Sam and Doc with a tired but welcoming smile.

Sam and Doc stepped forward, nodding respectfully. "Evening, ma'am. Evening, sir," Sam said, his voice low and steady. "We're here with Jack."

Doc added, "We've been through a lot together, and we'll be right here now too. He's family to us, and so are you."

Jack's parents exchanged a glance, their smiles faint but genuine. "It's good to see you boys," his mother whispered. "You've always been like sons to us."

Jack squeezed his mother's hand. "I wanted you to see them, Ma. They've been with me through everything, and I wouldn't be here without them."

The room settled into a quiet warmth, the three men standing by the beds, the sense of family stronger than any distance, any war, any time apart. Even during uncertainty, there was comfort here— in presence, in touch, in the unspoken bond that carried them through decades of life and loss.

Jack took a deep breath, letting the weight of the day ease just slightly. For now, they were together—and for now, that was enough.

As Jack, Sam, and Doc stood quietly by the beds, sharing soft words and small smiles, Jack noticed a subtle shift in his parents' expressions. His mother's eyes, so bright moments ago, flickered with confusion.

"Jack… have you boys… been here before?" she asked hesitantly, glancing at Sam and Doc.

Jack's father furrowed his brow, uncertainty in his gaze. "I… I don't remember. Were you… here with us already?"

Jack's stomach tightened. He had been so focused on bringing comfort and connection, yet now it seemed the passage of time—and perhaps illness—was erasing even the memories that mattered most.

Jack forced a gentle smile, squeezing his mother's hand. "Yes, Ma. We've all been here before," he said softly. "A long time ago. But we're here now, and that's what matters."

Sam stepped closer, resting a reassuring hand on Jack's shoulder. "Memory can fade," he said quietly. "But love… that doesn't. That sticks."

Doc nodded, his voice steady. "We've all been through enough to know what counts. Right now, this moment is what matters, Jack. You're here with them."

Jack looked at his parents again, their faces pale but peaceful, and he felt a mixture of sadness and determination. Time could steal memories, but it couldn't take the presence, the care, and the bond that still existed between them.

He leaned down, pressing a soft kiss on his mother's forehead. "We'll make the most of the time we have, Ma, Dad. I promise."

For a while, the room was quiet again, filled only with the faint hum of the hospital and the unspoken understanding between the three men and the parents they loved—fragile, fleeting, but fiercely held.

Jack stepped back slightly, letting Sam and Doc linger by the beds. He gazed out the window at the fading light spilling across Millersville, the warm glow of evening casting long shadows over the town.

For a moment, he allowed himself to think about all the years he had drifted wars fought, places left behind, people lost along the way. And now, here he was, standing with the two men who had been through every battlefield with him, at the bedside of the parents who had raised him.

Time had a way of slipping through fingers, he realized. Memories could fade, faces could blur, but the moments they were making now—quiet, fragile, filled with love—were theirs to hold.

Jack let out a slow breath, feeling a weight lift just slightly. He would carry these minutes, these smiles, and even the fleeting confusion in his parents' eyes with him. They were reminders of what mattered, what endured.

Turning back to Sam and Doc, he nodded. "Let's stay a little while longer. Every second counts."

And so, they did—three brothers-in-arms, a family held together by love and loyalty, cherishing the simple, precious truth that some moments, though fleeting, were worth remembering forever.

Jack sat close to his parents, holding their hands gently. The room was quiet except for the soft hum of the hospital machinery, the fading light of evening spilling across the beds. Sam and Doc lingered nearby, their presence steady, offering silent support.

Jack's mother stirred slightly, her lips curling into a weak smile. "Jack… I'm… so proud of you," she whispered, her voice barely audible.

His father's hand squeezed his own. "You've done well, son. You've always done well," he murmured.

Jack leaned closer, pressing his forehead gently against his mother's. "I love you, Ma… Dad… I'll always be here," he said, his voice trembling with emotion.

For a long, quiet moment, they stayed like that, the bond between them unbroken. Then, as the sun dipped below the horizon, Jack felt the hands in his gently go still. His mother and father had passed peacefully, side by side, their faces serene.

Sam and Doc moved closer, offering quiet words of comfort, but Jack remained kneeling, taking in the finality, the weight, and the quiet grace of the moment.

Outside the window, the evening shadows deepened, but inside, the love and memories they shared lingered, a tether that would never fade. Jack sat with them, silent, present, and resolute, knowing that though they were gone, the bond they shared would endure forever.

Chapter 4: Restless Roots

The Mercer farm stretched out before them, bathed in the soft glow of late afternoon sun. Jack, Doc, and Sam drove down the familiar gravel lane in quiet reflection, the air thick with the scent of freshly turned earth and summer grass. The funeral had been a somber gathering, but it had also been a testament to the lives Jack's parents had touched—neighbors, friends, and townsfolk from every corner of Millersville had come to pay their respects.

"They really showed up for them," Sam said, breaking the silence. "Every face in that crowd knew them, respected them. It was… something to see."

Doc nodded, his hands resting lightly on the dashboard. "Your parents were pillars of this town, Jack. You could feel it in the way everyone stood there, the quiet reverence. They weren't just your parents—they were family to all of us."

Jack exhaled slowly, his gaze sweeping over the farm stead. "Yeah… they were. And now it's just us here again. Feels strange, coming back after everything."

The truck rolled to a stop in front of the farmhouse. Dust swirled lightly in the heat of the afternoon as the three men climbed down, boots crunching against the gravel. The house, barn, and fields had endured decades of seasons, just as Jack's memories had endured decades of wars and distance.

Sam glanced at Jack. "Feels like the right place to be for a while. Surrounded by what mattered."

Doc gave a small, approving nod. "Let's settle in, take stock, and maybe get some dinner. A long day like today… we deserve a little normalcy, even if it's fleeting."

Jack looked out over the fields one last time before stepping toward the porch. The Mercer farm had always been home—and for the first time in years, after loss and long journeys, it felt like it might still be.

The front door creaked softly as Jack pushed it open, and the familiar scent of aged wood and lingering summer warmth greeted them. Doc and Sam followed, boots thudding lightly on the worn wooden floors.

Jack set his bag down and glanced around the room. Dust motes danced in the fading light streaming through the windows, and the furniture—simple, sturdy, and familiar—stood as it always had. It was home.

Sam sank into a chair near the hearth, letting out a low whistle. "Still looks the same, Jack. Feels… good to be back."

Doc leaned against the doorframe, scanning the room. "Yeah. Feels like the war and everything that came after just… paused for a moment."

Jack smiled faintly, pouring a cup of coffee for himself and handing one to each of them. "It's quiet here now," he said, taking a sip. "After the funeral, seeing everyone from Millersville there… it reminded me how much my folks meant to people. And how much this farm, this town… meant to me."

Sam nodded, lifting his cup. "They raised a good one. You did them proud, Jack. All of us did what we could to be there, but they made you who you are."

Doc chuckled softly. "Five days of relentless PT on the farm, endless drills before Normandy… I'd say they made all of us something stronger, too."

Jack laughed quietly at the memory. "Yeah… I guess so. You remember that first week? I ran you both ragged with every drill I could think of."

Sam grinned. "We survived it somehow. Barely."

Doc shook his head, smiling wryly. "Barely is right. But it made us better for what was coming. You, me, Sam… we all came out of that stronger because of it."

For a while, they sipped their coffee in silence, the weight of the day and the funeral settling around them, softened by the comfort of the familiar farmhouse. Outside, the fields swayed gently in the evening breeze, and inside, three brothers-in-arms allowed themselves a rare moment of peace and reflection.

After finishing their coffee, Jack, Doc, and Sam moved through the farmhouse, pausing at familiar spots—the kitchen table where Jack had done homework as a boy, the worn rocking chair by the window where his mother would sit and hum, and the old wooden chest at the foot of his parents' bed that had held decades of family keepsakes.

Doc ran his fingers lightly along the edge of the chest. "This place… it's like stepping back in time. You can almost hear the years echoing through it."

Sam smiled softly, glancing around the kitchen. "And smells too. I swear, every corner here carries a memory. Jack, your folks raised you right. You can see it in everything—how the house holds together, how the farm runs, even the way you take care of people now."

Jack paused at the doorway to the barn, letting his eyes drift over the fields beyond. "This farm… it's more than land. It's where Mom and Dad taught me what matters—hard work, loyalty, family. And now… all of it feels like it's tied to their memory, more than ever."

They stepped outside, the sun lower now, casting long shadows across the fields. Jack led them down a familiar path to the barn, the warm scent of hay and earth filling the air.

Doc let out a quiet whistle. "Five days of relentless PT, Jack. You ran us through everything here before Normandy. Looking at it now… it's easy to forget how brutal you were."

Sam laughed, shaking his head. "Brutal, yes—but effective. Made us better soldiers, better men. We all owe some of that to this place—and to your folks for putting up with us.”

Jack's smile was tinged with sadness. “Yeah… they endured it all with patience, and they loved every one of us like we were their own. That's why I wanted you two to see them, to understand what they meant to me and to everyone else.”

For a moment, they stood in the fading light, three friends surrounded by fields, barns, and the enduring spirit of Jack's parents. Words weren't needed. Every corner, every smell, every shadow spoke of family, loyalty, and the fleeting beauty of moments preserved in memory.

Sam leaned against the fence post; eyes fixed on the rolling fields that stretched beyond the Mercer farm. “So… now that it's yours, Jack,” he said quietly, “what's the plan? Are you going to… stop drifting?”

Jack shifted, letting the weight of the land and the memories sink in. The farm smelled of earth and old wood, familiar and grounding. He hesitated, staring at the horizon where the sun was just kissing the treetops.

“I don't know, Sam,” he finally admitted. “Part of me… wants to stay. Make this place matter again. But another part…” He let his voice trail off, unsure if the other part would ever be satisfied staying in one place.

Sam nodded slowly, understanding without pushing. “Well, wherever you go, Jack… you know we'll be there. Always.”

Jack managed a small, grateful smile. "Yeah... I know."

That evening, Jack, Doc, and Sam sat on the Mercer porch, the warm glow of sunset spilling across the fields. Jack cradled a mug of coffee, watching the gentle sway of the corn in the breeze. For the first time in a long while, the world felt quiet.

Doc leaned back in his chair, a soft smile tugging at his lips. "You know... seeing this place again... it feels like time slows down here."

Sam nodded, letting his gaze drift to the horizon. "Yeah. Hard to believe everything we went through... and now it's just... calm."

Jack took a slow sip, letting the warmth spread through him. "It's nice," he admitted. "Almost makes you think you could... settle down, even if just for a little while."

The three of them sat in companionable silence, listening to the evening sounds of the farm—crickets chirping, a dog barking in the distance, the wind whispering through the trees. It was a rare, fleeting peace, one that reminded Jack of what he'd been fighting for all these years

Jack lingered on the porch long after Doc and Sam had gone inside, the chill of dusk creeping into his bones. The farm stretched before him, golden fields fading into shadow, silent except for the whisper of the wind. Ownership—it sounded simple when he said it out loud. *The farm is yours, Jack.* But the words felt heavier than he expected.

He traced the edges of the old barn with his eyes, remembered his father's hands guiding him over the fence rails, the long summer days spent sweating in the sun, the winters that seemed endless yet somehow full of warmth. Could he really stay here? Could he trade the pull of the world, the rush of movement, for a life of quiet soil and harvests?

The thought of settling down stirred something foreign in him: a longing for roots, for permanence. But beneath it, a deeper, restless part of him twitched, the part honed by battles, missions, and the endless motion of a life that refused to age at a normal pace. Staying put felt like surrender.

Jack swallowed hard, the weight of choice pressing on him. Could he honor his parents' legacy and still honor the man he had become? Or was the farm simply a place to pause, a temporary refuge before the world called him back into chaos once more?

He closed his eyes, listening to the wind through the trees, the quiet of a world at peace, and wondered if peace was something he could ever truly hold onto—or if it would always slip through his fingers like sand.

Jack opened his eyes and let them wander over the fields again, imagining what it would take to run the farm. The barn needed repairs to the roof shingles missing, the paint peeling in long, tired streaks. The fields would require careful tending to bring them back to productivity. Crops, livestock, the endless rhythm of planting and harvest—it was a life of labor, but a tangible one, grounded.

He pictured himself rising before dawn, the smell of earth and hay filling his lungs, the sun climbing over the horizon to warm his back, the sweat on his brow a reminder that his work had meaning. He could do it—he *could* stay—but for how long before restlessness clawed at him again. The farm demanded commitment, routine, a permanence he wasn't sure he was built for.

And yet, even as he imagined life here, his mind drifted back to that cold, sterile lab in Normandy. The smell of antiseptic and the harsh light that never softened, the rows of glass vials, and the German scientists with their precise, indifferent expressions. The experiment. The way it had changed him, twisted time around his body while leaving the world moving too quickly outside him. He had escaped death, but at a cost. How could he ever truly belong anywhere, when every decade would pass him by like a storm he could never outrun?

Jack thought about Doc and Sam, their steady presences, and a pang of longing for a life that was simple, anchored, safe. Could he trade the thrill of the unknown, the call to action he had answered so many times, for long days under the sun, tending to the same soil his family had worked for generations?

He leaned back in the chair, listening to the hum of crickets and the soft rustle of the corn, the first fireflies beginning to blink in the twilight. Perhaps he could find a balance—stay long enough to honor the farm yet leave if the world demanded it. But even as the thought comforted him, a flicker of doubt burned. Was he capable of choosing roots over the restless road, or was he drifting forever in his life?

The warm June evening settled around him, heavy with the scent of earth and growing crops. Somewhere deep down, he knew that whatever choice he made, it wouldn't just define his time on the farm—it would define him.

The next morning, the sun spilled gold over the Mercer farm, the fields alive with the hum of summer. Jack, Doc, and Sam sat on the porch, mugs of coffee warming their hands, the scent of freshly turned earth mingling with birdsong.

Sam shook his head, staring out at the fields. "I still can't get Rizzo out of my mind… that day…"

Doc's jaw tightened, and he gave a slow, grim nod. "D-Day. They jumped just like we did. Rizzo and his team… they landed right in that field covered by a German MG-34. Didn't stand a chance. All six of them were cut down before they could even get out of their parachutes."

Jack's gaze drifted to the horizon, the warmth of the morning doing nothing to ease the chill in his memory. He could see it again—the chaos, the screaming, the bright flashes of gunfire, and the way the ground swallowed his friends in an instant. "I remember the sight… and the silence after. Nobody moved. Nobody lived."

Sam's hands clenched around his mug. "Damn… Rizzo was just a kid. Full of fire, like you said. Reckless, but damn, he had heart."

Doc stared down into his coffee, voice low. "We all lost too many that day. Rizzo and his men… just one more reminder of what war does."

Jack lifted his mug in a quiet, private salute. "To Rizzo. Gone too soon. But we remember."

The three men sat in silence, letting the summer morning surround them, even as the memory of that bloody Normandy field lingered like a shadow over the porch.

The distant hum of a car engine broke the quiet, pulling Jack, Doc, and Sam out of their thoughts. A sleek sedan rolled up the driveway, windows down, laughter drifting from inside.

Lisa stepped out first, her smile bright in the morning sun, followed by her husband, John—tall and easygoing—carrying a cooler and a few baskets. "Well, look at you three," Lisa called, waving as she approached the porch. "Thought we'd come by and see how the new owner of the farm is holding up."

Jack set down his mug, managing a small smile despite the heaviness still lingering from their conversation. "Morning, Lisa. Morning, John. Didn't expect company this early."

John tipped an imaginary hat. "Couldn't resist. Figured we'd check on the legend himself." He clapped Jack on the shoulder, a friendly weight that reminded him of camaraderie without the ghosts of war.

Lisa laughed, glancing around at the three men. "Looks like the farm is in good hands… or at least someone's trying."

Jack took a deep breath, the present wrapping around him like a bomb. The warmth of June, the sight of friends smiling, the gentle rhythm of life on the farm—it all contrasted sharply with the cold memories of Normandy. For a moment, he could almost imagine that he might stay, that life could be simpler, grounded, and whole.

Doc sipped his coffee and gave him a small grin. "Well, since you're here… grab a chair. We were just reminiscing about the old days."

Lisa raised an eyebrow. "Oh, I'm sure those stories are full of adventure… and a little trouble."

Jack chuckled softly, the sound catching the morning light. "Yeah… a little of both."

For a moment, the shadow of the past seemed to lift, leaving only the warmth of June, the hum of life around the farm, and the quiet comfort of friends by his side.

By late morning, Lisa and John had waved goodbye, their laughter fading down the driveway as they drove off. The porch felt quieter once more, the hum of the summer farm settled back into its natural rhythm. Jack, Doc, and Sam lingered over their mugs, the weight of memory and conversation still settling between them.

The radio, which had been playing softly in the background, suddenly cut through the comfortable silence with a sharper tone. The announcer's voice, urgent and clipped, grabbed their attention.

"—special report—conflicting reports coming in from the Korean Peninsula. Early this morning, North Korean forces crossed the 38th parallel and launched attacks against South Korea. American forces are being mobilized, but the situation is developing rapidly…"

Jack froze mid-sip, the coffee mug slipping slightly in his hands. Doc straightened, brow furrowed, and Sam muttered under his breath, a familiar tension creeping in.

The warm June morning, the smell of earth and growing crops, the quiet peace of the farm—it all seemed to hang suspended in that moment, fragile and fleeting. Outside, the fields swayed lazily in the breeze, but inside, the world had shifted again, and Jack felt the old pull stirring in his chest—the call to action, the restless tug that had defined him for so long.

The radio continued, words blending into a backdrop of unease, and Jack stared out over the horizon, knowing that whatever choice he made about the farm, the world beyond it was already moving fast, waiting for him to decide whether to stay or drift once more.

Chapter 5: A Life Out of Step

Jack, Sam, and Doc remained on the porch, frozen in place, the radio's urgent bulletin still echoing in their ears. The June morning had felt peaceful just moments ago, but now the warmth of the sun seemed almost deceptive, masking a world suddenly spinning out of control.

Jack set his mug down slowly, hands still trembling slightly. "North Korea… they actually did it," he murmured, his voice low, heavy with disbelief. "They crossed the 38th parallel… and hit South Korea."

Sam ran a hand through his hair, staring off into the fields as if the answer might be hidden in the waving corn. "Damn… it's happening fast. Just like that."

Doc leaned forward, elbows on his knees, jaw tight. "Yeah… and we're just sitting here. It feels unreal, but it's real. They're already moving. American forces… God, we might be headed back in sooner than anyone thought."

Jack's gaze drifted to the horizon, the farm stretching out in golden fields, serene and timeless. But the memory of Normandy, Rizzo, and the experiment that had made him something other than normal surged forward, reminding him that the world beyond these fields had a pull he couldn't ignore.

The radio continued its rapid-fire reporting, names, locations, and troop movements spilling over each other. Jack's jaw tightened. For a moment, the farm seemed like a fragile bubble, one that might not hold for long.

He looked at Doc and Sam, two constants in a life defined by chaos. "We've got choices to make… and fast," he said quietly, a tension settling deep in his chest. "Do we stay, or… do we answer the call again?"

The three men sat in heavy silence, the June morning now charged with an unmistakable sense of urgency.

Doc cleared his throat, breaking the tense silence. "I'm out," he said firmly. "The Army's already labeled me 'not combat effective.' Too much time as a POW back in the war. They won't touch me."

Sam shook his head, leaning back in his chair. "I'm not going either. The farm, my family… that's my duty. I'll protect it, but I won't be marching off to some foreign war."

Both men turned their eyes toward Jack, expectant, calm but steady. Sam leaned forward slightly. "Jack… now that this farm is yours, you'll be exempt. Head of the place, means you're needed here. You're safe."

Doc raised an eyebrow, his voice a bit sharper this time. "Safe? Don't kid yourself. If you don't volunteer, they could draft you anyway. They won't know your real age, Jack… you still look twenty-three. You're walking a thin line."

Jack's chest tightened. He stared down at his mug, feeling the weight of their words press in. The farm, the peace, the chance to finally feel rooted, it all hung in the balance. And yet, Doc's warning had stirred the old truth he'd carried for decades, a secret too heavy to leave unsaid any longer.

He took a deep breath, finally meeting their eyes. "There's something I need to tell you both," he began, voice low but steady. "Something you're going to need to understand before we make any decisions."

Doc and Sam exchanged glances, curiosity and concern flickering across their faces. Jack's heartbeat faster. The moment had come. He couldn't hide it any longer.

Jack took a deep breath and fixed his gaze on Sam. "Do you remember… back in Holland, when you asked me what the Germans did to me?"

Sam nodded slowly. "Yeah… I remember. You never said."

Jack swallowed hard. "There's a reason I never did. You need to know now." He paused, letting the words settle before he continued. "It started with the shots… a new serum they were testing. Then came the physical exertion. Hours, days… sometimes no rest at all. And more shots. Repeatedly, for three weeks straight. Every time, I was pushed beyond anything I thought my body could take."

Sam's hand tightened around the mug he was holding, his jaw clenched. Doc leaned forward, eyes narrowing, gripping the edge of his chair.

Jack's voice wavered slightly, but he pressed on. "The doctor… he didn't want to do it. He didn't like it, but he was being forced by the Germans. Everything they made me endure… he hated it, but he couldn't stop it. Until one day, I convinced him to help me escape. He risked his life for me, for a chance I had to take, to get out of that place."

Sam's face went pale, his fingers tightening around his mug so hard it rattled. "You… survived all that?"

Jack's lips pressed together. "I did. But it changed me, Sam. Not just my body… my life. The serum slowed my aging… ten years outside is one year for me. I've lived decades while the world barely noticed. That's why I look like I'm twenty-three. That's why I can't just… disappear into the farm forever. The army could call me, and they won't know how old I really am."

Doc ran a hand down his face, voice quiet but firm. "Jack… you should've told us sooner. That… that explains everything."

Jack leaned back in his chair; the weight of the secret finally lifted just a fraction. "I had to know I could trust you both first. Now you know. I don't have the luxury of hiding anymore. Whatever choice I make… it's got to be real. No half-measures."

Sam exhaled slowly, shaking his head. "Damn… Jack. That's… hell. I can't imagine what you went through."

Jack met his gaze. "You don't have to imagine. You'll see the world I've lived in, and why staying here… or leaving… isn't simple."

The three men sat in silence, the June sun warm on the porch, but the weight of Jack's confession hanging between them like a storm cloud ready to break.

Doc leaned back in his chair, staring out over the fields, his hands steepled in thought. "Jack… I understand it now. You're not just some kid who looks young. You've been through hell the rest of us can't even imagine. That… that changes everything."

Sam ran a hand over his face, shaking his head. "I mean… damn, Jack. You survived what no one should have survived. And these six years… you've been out there moving through the world while the rest of us… we age and live normally." He looked at Jack with a mix of awe and concern. "No wonder you've always been… different. But it also makes this draft thing complicated. I get why Doc warned you."

Jack stared at his hands, gripping the edges of the porch railing. The warm June air felt almost heavy, pressing down with the weight of choice. "I thought the farm might finally be a place I could stop, even for a little while. A place to be normal… or at least as normal as I can be. But hearing about North Korea, the way the world's still moving… I realize normal isn't in my cards."

Doc's voice was quieter now, but steady. "Jack… you don't have to go if you don't want to.

But you also can't pretend the world won't come knocking. And with what you've got, what you *are*, if they call… you'll have no choice unless you hide."

Sam placed a hand on Jack's shoulder. "We're not telling you what to do, brother. That's your choice. But whatever you pick, know this: we've got your back. Always. The farm can wait… or it can be your anchor. It's up to you."

Jack looked out over the fields, the sun glinting off the corn, the summer breeze carrying the scent of earth and promise. He thought of Rizzo and his team, the experiment, and the six years that had passed since—years in which he'd moved through life while the world outside aged normally. He thought of the farm, of Doc and Sam, of the quiet life he might grasp if he stayed.

And then he felt it—the undeniable pull, the stir of duty, of restlessness, of the life he'd been living long before the farm existed.

"I… I can't ignore it," he said finally, voice low but resolute. "If they need me… I can't just sit here. Not while the world is moving like this. But I won't leave without planning… without making sure the farm—and you two—are taken care of."

Doc and Sam exchanged a look, understanding in their eyes. "Then we'll help," Doc said quietly. "Whatever it takes."

Sam nodded. "Yeah. No matter what comes, we face it together."

Jack let out a long breath, feeling the weight of decision settling over him, heavy but right. The summer morning carried on, the farm alive and peaceful, but Jack knew—this calm was temporary. The world had already begun moving again, and so must he.

Jack took a deep breath, meeting Doc and Sam's eyes. "There's one more thing," he said, his voice steady but heavy. "I don't want anyone—*anyone*—to know about my… condition."

Doc frowned, leaning forward. "Your condition?"

Jack shook his head. "My aging. The serum… what the Germans did. If it was to get out, I'd be a lab rat for the government before you know it. They'd want to study me, experiment on me again, control me. I won't let that happen."

Sam's expression darkened, concern etched across his face. "Jack… that's… a lot. But I get it. You don't owe anyone that. Not after everything you've been through."

Doc's voice was low, thoughtful. "We'll keep your secret. Nobody will know. But… that also means you've got to be careful. Every choice you make from here on out… it matters."

Jack nodded, the weight of both truth and secrecy settled firmly on his shoulders. "I know. That's why I must be the one to decide. I can't let the world—or anyone else—take that choice from me."

Sam placed a hand on his shoulder. "Then we keep it between us. Always."

Jack let out a slow breath, feeling a measure of relief. At least here, with Doc and Sam, he could be himself. And for now, that was enough.

Jack leaned back in his chair, staring at Doc with a mixture of hope and hesitation. "Doc… you're still in the Army, right? Working in personnel management?"

Doc nodded slowly, eyebrows raised. "Yeah… what's on your mind?"

Jack's voice lowered, cautious. "Is there any way… any way you could get me back in the Army as a twenty-three-year-old? Make me reappear somehow… even if it means creating a new record, a new birth certificate… something so nobody asks questions?"

Doc hesitated, lips pressed together, clearly weighing the risks. "That's… risky, Jack. It's not exactly standard procedure. But…" He leaned closer, eyes locking with Jack's. "I know enough about the system, enough to pull strings quietly. With the right paperwork and some careful maneuvering, it might be possible. We'd have to make it airtight—no loose ends, no questions. It wouldn't be easy, but…"

Jack let out a slow breath, relief and apprehension mingling. "I must do it, Doc. The farm… it's important. But if the world calls, I can't just hide here. I need a way to answer when it comes."

Doc nodded, his expression was grave but resolute. "Alright. I'll investigate it. If anyone can make you disappear into the system and reappear as a twenty-three-year-old without raising alarms… I can."

Sam shook his head, half in disbelief, half in admiration. "Jack… only you could get yourself into something like this. But I guess if anyone's up for it, it's Doc."

Jack's eyes swept over the fields, the warm June breeze brushing against him, and he knew the choice was already setting itself into motion. The farm could wait… but soon, he would have to move.

By mid-morning, Doc had left the farm, heading into Millersville with a purpose. Phones rang, wires were sent, and discreet conversations were held in hushed offices and quiet corners. The hours passed in a blur of logistics, signatures, and careful maneuvering—everything had to be perfect, airtight, and invisible to anyone who might question Jack's sudden "reemergence."

When Doc returned to the farm that afternoon, he carried a worn folder under his arm and an expression that mixed pride, concern, and a hint of disbelief. Jack, Sam, and the farm seemed to fade from his mind as he sank into the porch chair, letting out a low breath.

"I did it," Doc said finally, voice steady but tense. "The paperwork's clean. You're… officially back in the Army. Assigned to the Ranger Infantry Company as a new incoming SSG." He let that sink in, watching Jack process the information.

Jack's eyes widened slightly, heart beginning to race. "Rangers…?"

Doc nodded. "Yeah. And it gets tighter. They've been alerted to deploy to Korea in July. They don't know your… special situation. All they see is a capable twenty-three-year-old SSG ready to join their ranks."

Sam whistled low, shaking his head. "Jack… you've got six years under your belt already. They're going to get more than they bargained for."

Jack leaned back, gripping the porch railing, the weight of the news settling like a stone in his chest. The farm, the quiet, the life he'd imagined for a while—already slipping further away. Yet a familiar pull stirred inside him: duty, action, and the restless current that had carried him through so many years of war and survival.

Doc gave a small, wry smile. "You wanted a way to answer the call… well, here it comes."

Jack's gaze drifted to the fields stretching out before him, the June sunlight warmed his shoulders. The world had started moving again, and he had no choice but to move with it.

Later that afternoon, as the heat of the June sun settled across the fields, Jack leaned back in his chair, his mind still buzzing with the news Doc had delivered. He watched Doc and Sam finishing their coffee, then cleared his throat.

"I've been thinking," Jack began carefully, "while I'm away… I don't want the farm to fall apart. I'm considering asking Lisa and John to run it for me."

Sam raised an eyebrow. "Lisa and John?"

Jack nodded. "Yeah. John knows his way around a farm—he's been a farmhand on Lisa's dad's place for years. He's reliable, hardworking… I trust him to keep things running smoothly while I'm gone."

Doc leaned back, rubbing his chin. "That makes sense. Someone the farm can depend on. And you don't want it all left to chance while you're overseas."

Sam let out a low whistle, but his expression softened. "I guess that works. At least you know it's in good hands. And Lisa… she's got a head for things too. Between the two of them, they'll keep the place afloat."

Jack's gaze drifted to the fields, feeling a small measure of relief. "I hate to leave it, but… it's better than letting it sit unattended. I want to know that when I come back, the farm's still my home, still ours."

Doc nodded firmly. "Then we make it happen. Talk to them, lay out what you need done, and trust them to handle it. You've got bigger things ahead, Jack."

Sam clapped him on the shoulder. "Yeah… you've fought wars before, and you'll handle this one too. The farm will be fine."

Jack allowed himself a small smile. For a moment, the weight of upcoming deployment felt slightly lighter, knowing the farm would be in capable hands while he answered the call the world was making.

Late that afternoon, a familiar rumble announced Lisa and John's arrival. They stepped out of the car, smiles bright in the golden light, carrying a basket of fresh-baked bread and a couple of mason jars filled with sweet tea.

Jack rose from the porch, offering them a warm, but slightly tense, smile. "Glad you came by," he said, motioning for them to sit.

Lisa set the basket on the porch railing. "Thought we'd drop in, see how things are going. The farm looks… peaceful. But I can tell something's on your mind."

Jack nodded, taking a deep breath. "It is," he said quietly. "I wanted to talk to you both. I… I have some unfinished business I need to take care of."

John frowned slightly. "Unfinished business?"

Jack hesitated, choosing his words carefully. "Yes. Something I must deal with… alone. I can't really explain it right now, but it means I'll be gone for a while."

Lisa exchanged a glance with John, concern flickering in her eyes. "Where… where are you going?" she asked softly.

Jack shook his head. "I can't tell you. Not yet. But while I'm gone… I want you two to move into the house and run the farm. Treat it like it's your own. Take care of it, make it thrive. I trust you to do that."

John's eyes widened, then softened with understanding. "Move in? Run it like it's ours?"

Jack nodded. "Yes. I know John, you've got the experience. Lisa, you both make a great team. I just… need to know that when I come back, the farm is alive, running, and taken care of."

Lisa's voice was quiet, almost reverent. "We'll do it, Jack. We'll take care of it."

John smiled, giving Jack a firm nod. "You can count on us. The farm will be fine."

Jack let out a long breath, a mixture of relief and lingering tension settling over him. "Thank you… both of you. I trust you completely."

As they chatted briefly about chores and the upcoming summer tasks, Jack couldn't shake the pull of the world beyond the farm. The warm June evening seemed to hold its breath, the calm before the storm. And for now, at least, he could leave, knowing the farm—and those he cared about— would be safe in capable hands.

The morning sun glinted off the polished rails at the Millersville train station, the hum of early commuters mingling with the distant whistle of an approaching engine. Jack, Doc, and Sam stood together on the platform, luggage at their feet, each feeling the weight of what was about to begin.

Doc adjusted his hat and clapped Jack on the shoulder. "Take care of yourself, Mercer. And remember… keep your secret close. Nobody else can know what you are."

Jack nodded, gripping the brim of his own hat. "I will, Doc. And thank you… for everything. I wouldn't be ready for this if it weren't for you."

Doc gave a tight smile, then nodded toward the arriving train. "Duty never ends, does it?"

Sam stepped forward, giving Jack a firm shake of the hand. "And I've got my own farm to keep in Boonesboro. But you… you're walking a long road again. Don't forget us back here, alright?"

Jack's eyes softened. "I'll never forget. You both have been my anchor."

As the train hissed and groaned, ready to depart, Doc leaned closer to Jack. "Don't forget—when you get to Fort Ord, they'll have a full issue of gear waiting for you. Uniforms, kit, the works. Once you're squared away there, you'll need to make your way to Travis Air Force Base. Be on that flight to Japan by July 5th, 1950. No excuses."

Jack gave a firm nod. "I won't miss it. You have my word."

The whistle of the train grew louder, and Sam clapped him on the back one last time before stepping toward his car parked near the station. Jack watched as Doc and Sam left, the familiar weight of separation settling in his chest.

With the train pulling into the station, Jack hoisted his bag over his shoulder, taking one last look at Millersville, the farm, and the friends who had anchored him. Then, with a final deep breath, he stepped aboard, setting off on the long journey to Japan and his new assignment with the 5th Ranger Infantry Regiment. The adventure, danger, and uncertainty that had always been a part of him waited just beyond the horizon, while the memory of home lingered like a quiet promise in his mind.

Part 2: Paths Converge

Chapter 6: The Weight of Time

The taxi rolled to a halt outside the main gate of Fort Ord. Jack stepped out into the California sun, the ocean breeze carrying faint salt over the dry hills. He handed the driver a few bills, then adjusted his stride toward the gate.

An MP flagged him down, looked over his papers, and scanned the orders clipped to the top. "Staff Sergeant Jack Mercer… age twenty-three." The MP glanced at him again but didn't question it, just stamped the orders and waved him through.

Inside the post, the hum of Army life surrounded him. Cadences carried over the asphalt, boots thundered on parade grounds, and the clang of rifles being racked echoed from the ranges. Six years had passed since Normandy, but the rhythm was the same, unchanged, relentless.

Jack walked past rows of barracks toward the Supply building, empty-handed. He wasn't here to turn in old gear; he had nothing left from the war. This time he was here to start over. A new issue waited for him: uniforms, boots, field pack, weapons, everything he'd need to step back into combat.

He paused for a moment outside the building, watching a line of recruits file past, laughing and shoving one another like boys. Jack wasn't one of them, not anymore, but he was on paper. Twenty-three years old. A Ranger Staff Sergeant. Ready for Korea.

With a steady breath, he pushed through the doors into Supply. Time to become that soldier again.

Inside the Supply building, the air was thick with dust, canvas, and the sharp tang of machine oil. Stacks of crates lined the walls; each stamped with block letters—U.S. ARMY PROPERTY. Behind a battered counter, a stocky man in khaki leaned on his elbows, a pencil tucked behind one ear. His hair was already gray at the temples; his jaw shadowed with permanent stubble.

He looked up as Jack entered, noting the paperwork. "Mercer, huh? Twenty-three, Staff Sergeant… You must've impressed somebody to pull those stripes this early." His tone wasn't mocking, just curious.

Jack forced a small shrug. "I've been around long enough."

The supply sergeant gave him a sharper look, then cracked a dry smile. "Name's Master Sergeant Callahan. I knew a Halpern once— dogface infantry, tough son of a gun. We both got nabbed in Normandy after the drop went bad. Krauts rounded up half our stick and shipped us off before we'd even fired a shot. You wouldn't happen to know him, would you?"

Jack's chest tightened for a moment. "Doc Halpern?"

Callahan's smile widened. "That's the one. Hell, we called him 'Doc' even though he wasn't a medic. Always patching up the boys with what little he could scrounge, always looking out for the others. Kept a lot of us breathing in that hellhole. If it weren't for him, I wouldn't be standing behind this counter." He leaned back, folding his arms. "Halpern told me he might be sending a man my way. Didn't say much about him, just that I should make sure he was squared away, no questions asked. Guess that's you."

Jack met the sergeant's eyes, steady. "That's me."

Callahan gave a slow nod, then reached for a stack of gear. "Well, Mercer, whatever your story is, you've got good friends in the right places. Let's get you kitted out."

For the next half hour, Jack was fitted head-to-toe: fatigues, boots, web gear, field pack, helmet. Callahan tossed each piece across the counter with the casual ease of a man who'd been issuing war for years. As Jack tried on the uniform, Callahan gave a short grunt of approval.

"Fits like you were born for it. Rangers will put it to use soon enough."

Jack tightened the laces on his boots, the weight of the new gear settled on his shoulders like an old, familiar burden. For the first time since Normandy, he felt the Army wrapping itself around him again shaping him, owning him.

Then Callahan leaned under the counter and hauled up a small crate stamped with fresh stencils: *TEST – NOT FOR GENERAL ISSUE.* He pried it open and pulled out a selection of new gear: a lightweight M1A1949 rucksack, reinforced combat suspenders, a sturdier field belt, and a prototype flak vest designed to absorb shrapnel. "Tell you what, Mercer. We've been issued some prototype gear—new harness system, lighter pack, even that vest there. They're not officially in circulation yet, but I figure a Ranger heading into the fire might as well give 'em a shakedown. You want in?"

Jack ran a hand over the harness and flak vest, feeling the weight and texture. "If it'll hold up under fire, I'll take it."

Callahan smirked. "That's what I figured." He started swapping out pieces, his movements efficient. "Consider yourself a field tester."

When Jack had finished strapping on the new gear, Callahan's expression sobered. He leaned on the counter, voice dropping low. "Word is this Korea thing… it ain't gonna be like Europe. No front lines, no gentleman's war. The Reds hit hard, and they're driving us south fast. They're saying it's already turning into a slaughterhouse."

Jack nodded once, silent, but his jaw tightened. He'd heard enough death warnings in his life to know when they were true.

Before heading for the door, Jack paused and extended his hand. "Thank you, Sergeant. I don't know what I'd have done without you helping me get squared away."

Callahan gripped his hand firmly, eyes sharp but warm. "Don't mention it, Mercer. You've got a long road ahead—just make sure you come back in one piece."

Jack gave a small nod, then turned, stepping out of the Supply building with the weight of his new gear and the burden of the war that awaited him.

Jack stepped out of the Supply building, the California sun glaring off the parade grounds. His new gear felt solid in his hands, packed neatly in a duffle bag slung over his shoulder. The fatigues Callahan had handed him were slightly faded, the fabric soft from prior use—a subtle touch that made him look like he'd been around the Army a while, not a fresh arrival.

He walked briskly toward the Transportation Office, the clack of his boots echoing on the concrete. Inside, the office was buzzing with activity: clerks filing papers, soldiers lining up with orders, the occasional whistle signaling incoming shipments of personnel and cargo.

Jack approached the counter, placing his orders in front of the clerk. "Staff Sergeant Jack Mercer. I need travel arrangements to Travis Air Force Base, California, for a flight to Japan—departure July 5th, 1950."

The clerk glanced down at the paperwork, typing swiftly into the terminal. "Everything checks out Sergeant. You'll get a train from Fort Ord to Travis, connecting to your flight. Orders are confirmed. You'll have transport details printed for you before the end of the day."

Jack nodded, satisfied. He tucked the slightly faded orders into his shirt pocket, the edges worn as if they'd been through previous campaigns.

The subtle aging of the fatigues, combined with his appearance, would help him blend in—nobody would question a seasoned Staff Sergeant, even if he was technically just back from civilian life.

As he left the Transportation Office, adjusting the straps of his duffle, Jack took a deep breath. The train to Travis was arranged, the flight booked, and soon he would be crossing the Pacific again. Japan awaited—and with it, the unknown dangers of a new war.

With four hours to spare before his train to Travis Air Force Base, Jack stepped out of the Transportation Office, the slightly faded fatigues Callahan had given him settling comfortably against his shoulders. He adjusted the duffle carrying his prototype gear and started walking across the post, taking in the familiar sights and sounds of Army life, the drill of marching soldiers, the bark of NCOs, the distant clang of rifles being cleaned on the ranges.

His route took him past the barracks and toward the chow hall. The morning sun was warm, and the steady rhythm of his boots on the pavement helped clear his mind. He thought about the upcoming flight to Japan and the uncertain war waiting for him in Korea, trying to steel himself for the unknown.

Inside the chow hall, the air smelled of hot coffee, grease, and the faint tang of metal from trays clattering on tables. Jack slid into an empty bench, setting down his duffle and digging into his breakfast—scrambled eggs, cold hash browns, and thick slices of toast.

A young sergeant, probably fresh out of training, slid onto the bench across from him, tray in hand, eyes bright with excitement. "Staff Sergeant Mercer, right? Heard you're headed to Korea. First deployment?"

Jack shook his head slightly. "No. I've been around."

The sergeant's grin widened. "That's awesome! I've been itching to get over there. Heard the action's going to be intense. Can't wait to get my hands on some real combat!"

Jack took a bite of toast, chewing slowly. He studied the young man—a fresh face, muscles tense with excitement, uniform crisp, eyes gleaming with untested confidence. He could see the enthusiasm, the thrill of heading into war, but he also saw the blind side of it.

"You're looking forward to it, huh?" Jack asked carefully.

"Absolutely! It's what we train for. I've been waiting for this my whole career," the sergeant said, leaning forward, voice full of energy. "Nothing beats getting in the fight."

Jack nodded once, letting his gaze drift across the hall. He remembered Normandy, the chaos of the drop, the sound of machine guns mowing down friends before they could even hit the ground. He thought of Doc and Sam, of the men who didn't make it back, and of the experiment that had kept him alive when all others had fallen.

"Just remember," Jack said slowly, setting down his fork, "war isn't like training. It isn't a game, and it isn't a parade. You won't get to pick the fight, and it doesn't matter how ready you think you are."

The young sergeant paused, confusion flickering across his face. "Uh… yeah, I guess I… I'll keep that in mind."

Jack offered a thin, knowing smile. "You will. Just make sure you survive long enough to see it."

The sergeant nodded nervously and went back to his meal, quieter now, chewing thoughtfully. Jack returned to his own breakfast, silent, letting the noise of the chow hall wash over him. He'd seen enough combat to know excitement didn't keep a man alive—and Korea was going to be unforgiving.

Four hours later, the sun had climbed higher, and the morning heat pressed against the concrete of Fort Ord's train platform. Jack adjusted the strap of his duffle, feeling the familiar weight of his prototype gear packed inside. The slightly faded fatigues Callahan had given him helped him blend in with the other soldiers moving about—seasoned enough to look like he belonged, though he knew the truth was far stranger.

The distant whistle of the approaching train echoed across the platform. Soldiers shuffled onto the wooden boards, duffels and packs in hand, some joking and laughing, others quiet and tense like him.

The train's conductor barked orders, and soldiers began boarding in orderly lines. Jack stepped forward, showing his orders to the clerk, then slid into his seat as the car door clanged shut behind him. The train began to move, and the countryside blurred past, the hum of wheels over rails marking the start of another chapter in a life that was anything but ordinary.

He stared out the window, watching Fort Ord shrink in the distance. This wasn't just another deployment. This was Korea—a new war, faster, deadlier, and far less predictable than Europe had been. Jack tightened his grip on the duffle, feeling the weight of both his gear and the burden of what was coming.

The train rolled steadily through the countryside, the fields and forests of California sliding past in a patchwork of green and gold. Jack pressed his forehead lightly against the cool glass of the window, letting the rhythmic sway of the train carry him forward. Outside, small farmhouses and winding roads gave way to stretches of empty farmland, the occasional herd of cattle grazing lazily.

But Jack's mind wasn't on the scenery. The gentle motion and the open view triggered memories he could never forget. Normandy. The drop. The morning haze that had masked the chaos waiting below. He saw Rizzo and his team, their chutes caught in the wind, twisting above a field.

Then the machine guns opened. MG 34 fire ripped through the sky and the field below, mowing down six men before they even hit the ground. Jack could still hear the rattle, the screams, the visceral panic of men realizing there was no escape. He remembered trying to move, trying to help, but being too far, too helpless.

He blinked, and the California fields replaced the French countryside—but the weight of that memory pressed against his chest just as heavily as it had on D-Day. Rizzo and the others weren't coming back. Not Rizzo, not his team. Not the men who had gone before him, who had never had a chance to see the war's end.

Jack exhaled slowly, forcing the past to the edges of his mind. He tightened his grip on the duffle. He had survived, carrying more than just his gear with him. And now, heading toward Japan and a new war in Korea, he knew the stakes were no less deadly.

Outside, the landscape continued to change rolling hills, small towns, and glimmers of the coast—but Jack's eyes remained distant, fixed on the past even as the train carried him forward into the unknown.

The California hills blurred past, but Jack's mind stayed in Normandy for a few more heartbeats, the memory of Rizzo and his team seared into his consciousness. He could still see the chaos, feel the panic, hear the staccato rattle of the

MG 34. Survival had been cruel, leaving him alive while others fell.

Now, Korea is awaiting. He knew it would not be like Europe. The reports were already trickling in: North Korean forces pushing hard, American units being forced south, civilians caught in the crossfire. Jack had seen combat before, but this would be different—faster, more brutal, and with enemies who didn't follow any rules.

He glanced at his reflection in the train window. Even with his slightly faded fatigues, he still looked 23—youthful enough to be drafted, old enough to know the cost of war. That was both a gift and a curse. He could deploy, fight, survive—but he carried a lifetime of experience no one else could understand.

Jack exhaled slowly, his fingers tightening on the duffle strap. He had unfinished business—not just with this new war, but with the memories that followed him from Normandy to Holland, from experiments he barely understood to the friends he had lost. Korea was another battlefield, another test of everything he'd endured.

The train's rhythmic clatter beneath him seemed almost like a heartbeat, carrying him forward toward the unknown. Jack closed his eyes briefly, steading himself. One way or another, he would meet what came—and he would survive.

Jack leaned back in his seat as the train carried him onward, the California landscape slipping past in a blur. Memories of Normandy and the fallen men he couldn't save pressed against him, a quiet reminder that survival carried its own weight. Korea awaited—unknown, dangerous, and unrelenting—but Jack had faced death before. He would face it again.

For now, all he could do was keep moving forward, one mile at a time, one mission at a time.

Chapter 7: Unspoken Bonds

The train rolled into the small station at Travis Air Force Base mid-afternoon, the clatter of wheels slowing as it approached the platform. Jack hoisted his duffle, feeling the familiar weight of his prototype gear, and stepped down onto the concrete. The air smelled faintly of oil, jet fuel, and dust—a strange mix that marked the edge of the Army and the beginning of something new.

Soldiers hustled past him, some in crisp new uniforms, others in fatigues softened by months of wear. Trucks and jeeps rumbled along the tarmac, while mechanics scurried beneath the wings of waiting transport planes. Jack scanned the activity, noting how organized yet chaotic the base seemed—every man and machine moving toward the same goal, the impending flight to Japan looming over them all.

He moved toward the operations office, where orders would be confirmed and boarding instructions issued. The faded fatigues Callahan had given him blended in perfectly; no one would question the seasoned look of a Staff Sergeant who had "been around," even if no one could guess the truth of his six years of unusual experience.

Inside the office, clerks shuffled papers and shouted instructions over the hum of fans and radios. Jack presented his orders, and a young lieutenant glanced at them quickly. "Staff Sergeant Mercer, all set. Your flight boards tomorrow morning. You'll be assigned seating on the transport, and your gear will be loaded directly onto the aircraft. Any questions?"

Jack shook his head. "No, sir. Thank you."

He stepped out of the office, adjusting the strap of his duffle. The base stretched before him, a living machine of preparation. Planes lined up on the tarmac, soldiers moving in precisely, practiced rhythms, engines humming as if aware of the urgency in the air.

Jack found a quiet corner near the flight line and settled down for a moment. He could hear the distant roar of aircraft engines, feel the vibration in the ground, and imagine the miles he would soon cover across the Pacific. Japan—and Korea— waited.

He took a deep breath, steadying himself. The past six years, the experiment, Normandy, the loss of friends—all of it had led him here. And now, with the world shifting once again, Jack Mercer was ready to step into the next chapter of a life that refused to follow the ordinary rules of time.

Jack moved along the flight line, observing the constant motion of soldiers and air crew. Boxes of supplies were being loaded onto aircraft, mechanics checked engines with meticulous precision, and orders were barked across the tarmac. The energy of the base was electric, a mixture of urgency and anticipation.

As he paused near a loading area, a young woman approached, clipboard in hand, issuing instructions to a group of nurses. She wore the crisp uniform of an Army nurse, her hair pinned neatly beneath her cap, and a calm authority in the way she moved.

"Excuse me," she said to a nearby officer, her voice steady. "We need to double-check the manifest before boarding. No one moves until it's verified."

Jack watched her for a moment, noting the confidence and grace of her every movement. Then she glanced up and their eyes met. She smiled—a small, warm curve of lips that made his chest tighten.

"Staff Sergeant Mercer?" she asked, tilting her head slightly.

Jack blinked. "Uh… yes, that's me."

"I'm Patricia Harrington. I'm headed to Japan as part of the Army hospital unit." Her accent carried the soft cadence of Iowa, just on the other side of the state from Millersville. "Looks like we're going to be traveling on the same transport."

Jack felt an unexpected surge of connection. He extended his hand, trying to mask his sudden awareness of her presence. "Jack Mercer. Pleasure to meet you."

She shook his hand, her grip firm but delicate. "Likewise. I don't suppose you've been to Japan before?"

"Not yet," Jack replied, keeping his tone casual, though his mind was racing. "First time."

She smiled again, and something unspoken passed between them—an instant familiarity, a spark of something rare. Jack knew immediately he was captivated. Beautiful, confident, and from almost the same corner of the world as him, she was unlike anyone he had met in years.

As they continued talking, Jack realized this wasn't just idle conversation. Patricia Harrington was the kind of woman who could hold your attention without trying, and he found himself wanting to know everything about her—even as the base's relentless activity and the looming deployment to Japan hung over them both.

Over the next hour, Jack found himself crossing paths with Patricia several times as the final preparations for the flight continued. She moved with the quiet efficiency of someone used to responsibility, checking manifests, directing nurses, and offering instructions with calm authority. Every time their eyes met, Jack felt a jolt a mix of admiration, fascination, and something deeper he couldn't name.

He fell into step beside her as they walked toward the boarding area. "So, Iowa, huh? That's not too far from home," he said, trying to keep the conversation light.

"Yes," Patricia replied with a soft laugh. "Not too far. I grew up just on the other side of the state from Millersville. Funny how life brings you places you'd never expect."

Jack smiled, thinking of the farm, Doc, Sam, and the years he had lived that seemed heavier than they looked. He wanted to tell her everything about himself—but he couldn't. Not yet. Not ever, perhaps.

As they walked, he noticed the way her hair caught the sunlight, the gentle tilt of her head when she listened, the sincerity in her expression. He felt himself drawn to her, a rare warmth breaking through the usual armor he carried.

But beneath it all, a shadow lingered. Jack knew the truth—he would age far slower than anyone around him. If life went as, it naturally did, Patricia would grow old, and he would remain young. She would live decades he could only watch from the sidelines. The thought was a silent ache, a caution he tucked deep inside, even as he laughed at one of her quiet jokes.

He forced himself to focus on the present. For now, they walked side by side, sharing small talk and smiles. He memorized the sound of her laughter, the tilt of her chin, the easy grace of her movements. He didn't allow himself to hope too much, but he knew that even a fleeting connection like this was worth holding onto.

The boarding call sounded in the distance, a sharp bark over the loudspeakers. Patricia glanced at him. "Looks like it's time."

Jack nodded, adjusting his duffle. "Time to see what's waiting for us across the Pacific."

As they walked toward the transport together, Jack kept that secret truth locked away— his unnaturally slow passage through time. And as much as he wanted to, he couldn't tell her that he would outlive her, that the years would treat him differently than anyone else he would ever meet.

For now, there was only this moment, and Jack decided to let it be enough.

Patricia adjusted the straps of her duffle as she walked toward the transport, her mind unsettled in a way she couldn't quite explain. There was something about Jack Mercer—something magnetic, a quiet intensity beneath his calm demeanor—that drew her attention. She found herself watching him more than she should, noticing the subtle lines in his face, the way he carried himself with the weight of experience far beyond his apparent age.

She shook her head, trying to dismiss the flutter of curiosity. Yet, every time he looked her way, a strange familiarity stirred in her chest. It was like a half-remembered dream, an echo of something deep in her past that refused to be ignored.

Patricia carried her own secret; one she could never reveal. Born in Germany in the early 1920s, she had been subjected to the same experiment that had marked Jack—a cruel sequence of injections and trials, designed to alter her body's perception of time. She had aged only one year every ten, a miracle that had saved her life but condemned her to solitude in plain sight. By the early 1930s, the work had ended, and she had escaped Germany in 1938, fleeing to Iowa where she lived quietly, blending into normal life while keeping her secret hidden.

Now, decades later, the weight of that secret pressed against her again. Here she was, thousands of miles from home, standing next to a man whose existence mirrored her own in ways no one else could understand. She felt an unspoken kinship with Jack, though she had no idea why. The pull toward him was undeniable, magnetic, and terrifying at the same time.

She swallowed hard, glancing away as the boarding call echoed across the tarmac. Jack was smiling at her, his gaze steady, calm, confident. She wanted to say something, anything but the truth of who she really was could never be shared. Not yet perhaps not ever.

Instead, she fell in step beside him, keeping her thoughts tightly guarded. And even though she wouldn't speak of it, even though the secret burned in her chest, Patricia felt a rare, fragile hope. That in a world full of ordinary people, she had finally found someone who, in some mysterious way, might understand.

The boarding call boomed again, louder this time, and Jack and Patricia moved toward the transport together. Soldiers shuffled past, their footsteps echoing against the metal ramp as they filed into the cabin. The hum of the engines and the scent of oil and fuel filled the air, mingling with the nervous energy of the men and women boarding.

Jack found two empty seats near the center of the cabin and motioned for Patricia to sit beside him. She hesitated for only a moment before settling in, adjusting her clipboard and bag on the floor.

For the first time since meeting, the tension between them softened. They were no longer just strangers on a airplane; they were two people about to cross the Pacific, sharing the anxiety of the unknown and the strange comfort of companionship.

"So," Jack began casually, "Iowa, huh? You've been an Army nurse long?"

Patricia smiled, a little wistful. "A few years now. But… I guess I've always wanted to help people. Even before I joined, I wanted to make a difference somehow." She paused, glancing at him, her tone growing softer. "Actually… I wasn't born in Iowa. I was born in Germany."

Jack's head turned slightly, curiosity sparking in his eyes. "Germany, huh? That's… quite a move. When did you come to the States?"

Patricia exhaled slowly, a small, nostalgic smile tugging at her lips. "1938. I escaped before… before the Nazis really started their craziness. It was… a long time ago. I've lived quietly in Iowa ever since."

Jack nodded, sensing her need to keep the story brief, yet he could feel the depth of what she had endured beneath the surface. "Iowa seems a world away from that," he said softly. "But I get it—sometimes life gives you a second chance, and you just take it."

She met his gaze, her eyes reflecting a mixture of gratitude and relief. "Exactly. And… well, it feels good to talk to someone who seems… not ordinary. You know? Someone who listens without pretending to judge."

Jack smiled faintly. "I know the feeling. Sometimes it's rare to find someone who can just… understand, even a little."

For a moment, the noise of the transport, the soldiers, and the engines faded into the background. In that small space, with the Pacific waiting ahead and war looming, Jack and Patricia found a fragile comfort in each other's presence, a tentative connection neither fully understood but both silently welcomed.

The transport doors closed with a metallic clang, and the engines began to hum louder, the vibration rolling through the floor beneath them. Jack and Patricia settled into their seats; the cabin filled with the low murmur of soldiers preparing for the flight ahead.

"So," Jack said, leaning slightly toward her, "do you ever get used to this—the constant moving, the uncertainty?"

Patricia chuckled softly, glancing out the small window at the sprawling tarmac below. "I suppose you learn to adapt. You find small things to hold onto while everything else keeps changing. That's part of why I wanted to become a nurse… something to help people, even in the chaos."

Jack nodded, letting her words settle. "I get that. Some things you just hold onto, no matter where life takes you."

A pause stretched between them, comfortable in its silence. They watched as soldiers moved about, stacking duffels and double-checking gear, the tension and energy of departure surrounding them. There was something in the quiet that felt private, a small island amid the rush.

"You seem… different," Patricia said suddenly, turning to look at him. "Not in a bad way. You carry yourself like… you've seen more than most. And not just the obvious kind of experience. Something deeper."

Jack smiled faintly, a hint of a shadow passing over his eyes. "Maybe. Some things you can't explain. Some things you just live with."

Her smile lingered, and Jack felt a rare warmth in his chest. He wanted to reach out, to share, but the weight of his secret, the truth of his life, of how time moved differently for him—kept his lips sealed.

Patricia, too, felt the pull, the sense of understanding without words. She wanted to speak of her own past, of the years she had lived differently than the world expected, but instinctively she remained silent. Some truths were too heavy, too strange to share—not yet.

For the remainder of the ride to the airstrip, they talked quietly about family, their lives in Iowa, and the mundane hopes of soldiers headed overseas. Laughter and easy smiles slipped through, the beginnings of a fragile trust. And though neither revealed the secret that bound them in ways they couldn't yet comprehend, an unspoken recognition lingered between them, a hint that they were more alike than either dared admit.

As the engines roared to full power and the transport began to taxi down the runway, both Jack and Patricia looked out at the horizon. The Pacific awaited, Japan beyond, and with it, the unknown dangers of war. Yet in the shared silence of the cabin, they found a small comfort: two souls quietly aligned, carrying unspoken burdens that the world could never understand.

The engines thundered, rattling the cabin as the transport gathered speed along the runway. Jack gripped the edge of his seat, eyes fixed on the horizon, feeling the ground fall away beneath them. The wind outside screamed against the fuselage, a reminder that nothing in this world stayed still for long.

Patricia adjusted her seat beside him, glancing briefly at him with that soft, reassuring smile. Jack offered a small smile in return, though inside, his mind was already hundreds of miles away—across the Pacific, toward Japan, and the unknown dangers waiting in Korea.

He thought of Normandy, of Rizzo and his team, of the years since that drop and the experiment that had kept him alive. The war ahead would be different, faster, and less predictable. Yet he carried all of it with him—the knowledge, the loss, the weight of survival—and he knew he was ready.

And then there was Patricia, sitting quietly beside him, her presence a strange, comforting anchor in the storm of his thoughts. He didn't know her secret, she didn't know his, but there was something unspoken that lingered between them. Recognition, a shared understanding that life had not treated either of them like everyone else.

The transport lifted off the runway, the sprawling base shrinking beneath them. Jack exhaled slowly, letting the roar of the engines fill the space between them. Ahead lay uncertainty, danger, and a war he could not yet imagine—but for this moment, there was a small thread of connection, fragile but undeniable.

Chapter 8: Anchors of Time

The transport shuddered as it descended over the Japanese coastline, the sprawling city below a patchwork of rooftops, roads, and distant mountains. Jack's eyes were fixed on the horizon, the Pacific behind them, and the anticipation of what awaited filling his chest with restless energy.

The transport's wheels screeched against the tarmac as it slowed to a stop at the Japanese airstrip. Soldiers surged forward, gathering their gear and heading toward the exit. Jack hoisted his duffle and stepped down, the humid air heavy with the scent of diesel and earth. Patricia followed close behind, adjusting her bag as they moved toward the in-processing building.

Inside, the room buzzed with activity—clerks stamping papers, officers calling out names, soldiers lining up for briefings. Jack and Patricia slipped into a quieter corner, letting the chaos of the room fade around them.

For a moment, they simply looked at each other. The war, the journey across the Pacific, and the unknown dangers awaiting them hung heavily in the air, but so did something else, something deeper neither had acknowledged until now.

Jack spoke first, his voice low. "I don't know… why, but I feel like I've known you longer than a few days. Like somehow… you're meant to be here."

Patricia's lips curved into a small, reflective smile. "I feel it too. There's something about you… it's like I can trust you without knowing why. And yet…" Her gaze dropped briefly. "I can't tell you everything about me. Not yet. Some things… some things are too strange to explain."

Jack nodded, understanding more than he could put into words. "I know the feeling. Some truths… some parts of life… you just keep locked away. Even from the people you care about most."

She leaned slightly closer, her voice barely above a whisper. "I left Germany before things… before it got bad. I've kept a lot hidden ever since, built a life where no one asks questions."

Jack's chest tightened. "That must have been difficuLT And yet you've made a life here, helping others, doing good in a world that doesn't always make sense."

Patricia nodded, her eyes searching his eyes. "And you… you carry something too. I can see it. You've been through things most people can't even imagine. And somehow, you survived. Somehow, you're here."

Jack exhaled slowly, letting the weight of it all settle. "It's been a long road. But being here now… seeing you… it feels like maybe there's something worth holding onto."

Her gaze softened, and for the first time, the tension of the unknown lifted slightly. "I'm glad you feel that way. Because… I think I do too. Even if the world around us doesn't make it easy."

They stayed in that quiet bubble as the rest of the room moved around them, their conversation weaving between shared understanding, unspoken fears, and a fragile hope. When the processing finally started to pull them into the official line, they lingered a moment longer, hands brushing briefly, hearts acknowledging the spark forming between them.

"We'll find each other again," Jack whispered.

Patricia nodded, her eyes glistening. "No matter what. Promise me, Jack."

"I promise," he said firmly, his voice steady.

They shared one last, lingering kiss before reluctantly pulling away, both aware that duty was calling and that the war outside this room was waiting.

Jack squared his shoulders, cast one final glance at Patricia, and moved toward the processing desks. Her presence stayed with him, a small but unshakable anchor amid the uncertainty of what lay ahead.

Jack moved through the processing line, but his mind wasn't on the papers or the clatter of stamps. It was on Patricia—on the way she had looked at him, the warmth in her eyes, the ease of the conversation that had felt impossibly rare in a world filled with orders, duty, and war.

He felt a pull toward her unlike anything he had known, a connection that went beyond words, beyond logic. And yet, a heavy weight settled in his chest. He couldn't be with her—not truly—not while he carried the secret that had defined him for years. The experiment, the altered aging, the knowledge that time moved differently for him than for everyone else. She deserved a normal life, a life he could never give.

The memory of Normandy struck next, vivid and painful. He saw Rizzo and his team jumping from the plane, the rush of wind, the sudden hail of bullets from the MG 34 that tore through the field. Six men, gone before they even hit the ground, their screams mingling with the chaos around him. He could feel helplessness all over again, the rage, the guilt, the sorrow that had never left him.

Shaking his head, Jack exhaled and tried to center himself, focusing on the present. The air in the processing building was thick with paper, ink, and the faint tang of sweat, but it grounded him. He squared his shoulders and pushed the memories to the back of his mind, forcing himself to breathe.

"Mercer! Staff Sergeant Mercer!"

The sharp call cut through his thoughts. Jack turned, his heart still tugged in two directions—toward duty and toward Patricia. The NCO's voice was insistent, carrying the authority of orders that could not be ignored.

Jack straightened, gripping his duffle tighter. The war waited, and he had no choice but to move forward.

Jack's gaze drifted back to Patricia, and for the first time in a long while, he allowed himself a fleeting thought of hope. Whatever came next, whatever battles waited on foreign soil, they were in this together, at least for now and that alone was something worth holding onto.

Jack followed the NCO across the base, the humid Japanese air clinging to his uniform. Soldiers bustled past in all directions, the energy of a war-ready staging area everywhere he looked. His mind still lingered on Patricia, but he forced himself to focus. Duty came first.

The NCO led him into a building marked for the Ranger company's temporary command post. Inside, the air was cooler, tinged with the scent of coffee and the sharp tang of bureaucracy. Officers and NCOs moved efficiently between desks, barking orders, signing manifests, and coordinating the next wave of deployments.

"Staff Sergeant Mercer, report to Captain Reynolds," the NCO barked, pointing toward a small office at the far end. "He's waiting for you. You'll be briefed on your assignment immediately."

Jack nodded, straightening his uniform and adjusting the straps of his pack. He walked with measured steps, aware of the weight of his gear and the gravity of the moment.

Captain Reynolds looked up as Jack entered, his eyes sharp and assessing. "Mercer, I've been told you're experienced, that you've seen combat," he said, scanning Jack's dossier. "The Ranger company isn't scheduled to arrive in Japan for another month. In the meantime, you'll be assigned to temporary duties here to assist with staging and preparation. Are you ready for this?"

Jack met the captain's gaze evenly. "Yes, sir. I'm ready."

"Good. You'll be integrated with other temporary personnel, issued your equipment, and briefed on operational procedures. Move quickly, there's no time to waste. When the Ranger company arrives, you'll join them and be ready for deployment immediately."

Jack swallowed, feeling the familiar adrenaline stir in his chest. The weight of responsibility pressed on him, but so did a thread of anticipation. He'd been trained for this, prepared for it, and yet the unknowns of what lay ahead made every step uncertain.

As he saluted and stepped out to begin in-processing and equipment issue, his thoughts flickered once again to Patricia. The memory of her eyes, her touch, the unspoken promise they had shared, stayed with him, a fragile anchor amid the chaos.

He had no time to dwell, no time to question what might come next. The war had already begun, and he had a role to play.

Jack stepped out of Captain Reynolds' office and was directed to the staging area for temporary personnel. Rows of lockers, weapon racks, and neatly stacked gear surrounded him. A sergeant approached, clipboard in hand, followed by a few younger soldiers.

"Staff Sergeant Mercer?" the sergeant asked, extending a hand. Actually, let's keep it simple—this is Sergeant Frank Davis. Just arrived, same as you. They've got us here until the Ranger company shows up."

Jack shook his hand firmly. "Jack Mercer. Nice to meet you."

Davis gave a quick nod. "Infantry, Normandy, the works. Thought I'd catch my breath in Japan before getting tossed into the next fight. Looks like we're in the same boat." He gave a wry smile, adjusting his duffle.

A few steps behind them, three younger soldiers eyed Jack and Davis curiously.

"That's Private First Class Tom Hill," Davis said, motioning to the tall, lanky kid. "Fresh out of Fort Benning. First overseas assignment. Eager, a little green."

"Good to meet you," Jack said, nodding to Hill, who straightened proudly.

"PFC Eddie Kim," Davis continued, motioning to a wiry man adjusting his helmet. "From California. Volunteered for overseas duty— figured Korea would be interesting." Kim gave a slight bow and a small grin.

"And PFC Jimmy Reilly," Davis finished, pointing to a freckled Irishman who looked nervous but determined. "He's from Boston. Joined last year. Family of cops, wanted to do something a little different."

Jack studied them all for a moment. Young, eager, untested in the crucible of combat. They reminded him a little of Rizzo's team before D-Day.

"Looks like we're going to be stuck together for a while," Jack said lightly. "Might as well make the best of it."

Davis chuckled. "Stick with us, we'll survive. And if we don't… at least we'll go down laughing."

Jack allowed himself a small smile, feeling the familiar rush of responsibility for others. He'd carry them through this waiting period and beyond if necessary.

As they moved toward the equipment area to get issued gear, Jack's mind drifted again to Patricia. He imagined her smile, the way her eyes lit up when they talked. He shook his head, forcing himself back to the present. There was work to do, and his temporary comrades depended on him— green or not.

The clang of metal and the smell of freshly oiled rifles filled the staging area as Jack, Davis, Hill, Kim, and Reilly were issued their gear. Backpacks, helmets, uniforms, and weapons were handed out in quick succession, each soldier checking the fit and function of their equipment.

Jack's eyes scanned the racks of rifles until he spotted what he was looking for. He approached the supply sergeant. "I'm going to need a paratrooper M1A1," he said firmly. "I'm very familiar with it from my previous service—handles better for me in the field."

The sergeant raised an eyebrow but handed over a sleek, slightly worn M1A1 carbine. Jack ran his hands over it, testing the balance and the trigger. "Perfect," he muttered. The familiar weight felt right in his hands, a connection to the battlefield he had survived before.

Once everyone had their gear, an NCO directed them to their billets. The room was small, utilitarian, with bunks lined up neatly and a single table pushed against the wall. The men stored their gear and settled in briefly before Jack stood, motioning for the others to follow.

"Alright," Jack said, his voice calm but commanding. "No one's ever heard of my workout plans before, but that's about to change. We're going to hit the training field and run a full PT session—cardio, strength, endurance—you name it. By the time the Ranger company arrives, we'll be in shape and ready to integrate immediately."

Davis, adjusting his duffle, grinned. "Never heard of them? Sounds interesting. Lead the way, Mercer."

Hill, Kim, and Reilly nodded, curiosity and apprehension mixing on their faces.

As the five of them stepped out into the humid Japanese morning, Jack slung his M1A1 across his back and glanced at the new team. They were fresh, untested, and eager—but they would follow him.

Jack inhaled deeply and told himself this would help. Focusing on the workout, on building their endurance and cohesion, would keep his mind off Patricia, keeping the ache of leaving her behind from consuming him. He led the group toward the training field, the first of what would be many grueling sessions, each push-up, sprint, and lift grounding him in the present.

Duty first. Preparation second. And Patricia… well, she would have to wait.

After the morning PT session, Jack wiped the sweat from his brow and slung his M1A1 over his shoulder. The grueling workout had done its job—his mind felt clearer, less tangled with the ache of missing Patricia—but he couldn't let the day end without seeing her.

He made his way to the base hospital, moving briskly through the corridors until he found the ward where she was assigned. The scent of antiseptic and the quiet hum of nurses and patients greeted him. He spotted her almost immediately, sitting at a desk, reviewing charts.

"Patricia," he said softly, stepping closer.

She looked up, her eyes lighting up the moment she saw him. A smile spread across her face that made his chest tighten. "Jack! I didn't think you'd have time to come by."

"I had to," he said, closing the distance between them. "Even a few minutes is worth it."

She stood, and he held out his hands. She took them, and they hugged, a warmth that chased away the tension of the past weeks. The embrace lingered, and then, almost naturally, it turned into a gentle, passionate kiss—one filled with unspoken promises and the recognition of a bond neither could deny.

They pulled back slightly, foreheads resting together. "I hate that we're here, in the middle of all this chaos," Patricia whispered.

Jack shook his head, a soft smile tugging his lips. "I hate it too. But seeing you—even for a few minutes—makes it bearable."

She tilted her head, brushing a strand of hair from her face. "Jack… do you ever worry about what comes next?"

He squeezed her hands. "Every day. But right now, we have this. We have each other, even if it's brief. That's enough to keep me going."

Patricia's smile softened, and she rested her hands on his chest. "I wish I could be with you out there."

Jack swallowed hard, hiding the weight of his secret behind a small, reassuring smile. "You will. Somehow, we'll find each other again. I promise."

They lingered a moment longer, savoring the connection, before Jack reluctantly let go. Duty awaited, and he had to return to the training field.

"I'll see you soon," he said, planting a quick, lingering kiss on her forehead.

Patricia watched him leave, her heart tight with longing but her spirit buoyed by the knowledge that Jack would return—and that the bond they shared was stronger than any distance or danger.

Jack walked out of the hospital, taking a deep breath. The workout, the new team, the upcoming deployment, all of it felt more manageable now. For a moment, he had seen her, held her, and reminded himself of what he was fighting for.

The afternoon sun was dipping low when Jack left the hospital, the faint scent of antiseptic still clinging to his uniform. Patricia's image lingered in his mind—the warmth of her hands, the soft smile, the quiet strength that drew him in like a magnet. He had no time to dwell, but the memory gave him something steady to hold onto amid the uncertainty ahead.

He walked back toward the billets, feeling the weight of his gear and the responsibilities waiting for him. The new team would be expecting him to get them ready for the next phase of his infamous—though so far unheard-of workout plans. He would push them hard, shape them, and keep their minds focused, just as he kept his own thoughts focused on survival and duty.

A small, private part of him ached, though knowing the connection he shared with Patricia could never be fully explored while he carried his secret, while the war stretched ahead. But he swallowed it down, letting it fuel him instead of slowing him.

Duty first. Preparation second. Love… he would carry that quietly, a secret anchor as vital as any weapon or piece of gear.

As the base quieted for the evening, Jack stepped into the billets, his M1A1 slung over his back, ready to take the new team through their first evening session. The day had been long, the hours heavy with thought and training, but he felt a quiet determination. Tomorrow, they would run harder, push further, and come closer to being ready for the Ranger company's arrival—and for the battles that lay ahead.

He paused at the doorway, looking back toward the hospital one last time. Patricia had given

him something that no war, no mission, no secret
could take away: hope. With that hope tucked safely
in his chest, he turned fully toward his team, ready
to lead.

Chapter 9: Forging Steel

The next month passed in a blur of motion, sweat, and discipline. Jack pushed the temporary team harder than they had ever been pushed before, and they responded in ways that surprised even him. Each morning began with grueling runs through the Japanese base, stretching muscles that ached in ways they had never ached before. Push-ups, pull-ups, sprints, and obstacle courses became routine, each exercise designed to build strength, endurance, and cohesion.

Afternoons were reserved for weapons training. Jack oversaw their handling of rifles, pistols, and grenades, emphasizing precision, speed, and safety. He demonstrated techniques he had honed during WWII, particularly with the paratrooper M1A1, showing the young soldiers how to maintain control and accuracy even under fatigue. Davis, Hill, Kim, and Reilly adapted quickly, their confidence growing with every drill.

Evenings were quieter but no less demanding. Jack used the time to review tactics, stress teamwork, and instill the mindset of paratroopers in his team. Though the Ranger company wouldn't arrive for another month, the temporary soldiers were beginning to feel like a unit, molded under his guidance and relentless energy.

Through it all, Jack kept his thoughts of Patricia tucked safely in the background. Each push-up, each lap, each fired round was a way to stay grounded, to channel the longing he could not voice. The rhythm of training gave him purpose, and the bond forming with his temporary team gave him the feeling of responsibility to match.

By the end of the month, they were leaner, faster, and more cohesive. They could run longer, lift heavier, and shoot straighter than when they arrived. Jack watched them, pride flickering beneath the surface of his stoic exterior. The Ranger company would arrive soon, and he would be ready to integrate this group seamlessly into a fighting unit.

The mornings were brutal, but each soldier began to find their rhythm.

Davis proved to be a natural leader beside Jack, pushing himself hard and quietly encouraging the younger soldiers. He had been through Normandy and other battles, and his presence gave the team a sense of credibility. During a long run along the perimeter of the base, he fell into step beside Hill and said, "Keep your eyes on the horizon, kid. The body can do more than the mind thinks." Hill nodded, panting, but inspired to keep pace.

Tom Hill had the most room to grow, but Jack noticed his determination. One afternoon during weapons training, Hill struggled with accuracy under fatigue. Jack walked over and demonstrated a stance adjustment and controlled breathing, then had Hill repeat the exercise. On the third try, Hill's shots were grouped almost perfectly. The pride on his young face mirrored the satisfaction Jack felt seeing his teaching take hold.

Eddie Kim had a sharp mind for tactical thinking. While running obstacle courses, he suggested slight modifications to formations that improved efficiency. Jack recognized the spark and quietly encouraged him to think like a leader. "Observing is just as important as doing," Jack told him one morning. "Notice everything, because it could save your life one day."

Jimmy Reilly was brash and impulsive, often trying to rush through drills. Jack pulled him aside during a grenade exercise, demonstrating the importance of patience and timing. Reilly's improvement over the next week was dramatic, and he began to gain confidence, understanding that raw enthusiasm needed guidance to become effective.

Jack himself pushed harder than anyone, demonstrating techniques, running alongside them, and correcting mistakes. He made sure each soldier felt the discipline without feeling crushed. The work was relentless, but there was humor too—grunts exchanged joking banter between sprints, and small victories were celebrated quietly but sincerely.

Weapon training sessions became another arena for bonding. Jack showed the team proper maintenance of the rifles, handling drills, and controlled bursts. His paratrooper M1A1 became a reference point; the young soldiers watched him work with the ease of someone who had lived with the weapon in the crucible of war. Each day, their accuracy improved, their confidence with firearms growing steadily.

Through it all, Jack kept his thoughts of Patricia tucked safely away. Her memory fueled him quietly, a reminder of life waiting beyond the rigors of duty. He imagined her smile, and the small stolen moments they had shared, letting that warmth sustain him during the toughest drills.

By the end of the month, the team had transformed. They moved with coordination, confidence, and trust—traits essential for any paratrooper unit. Jack stood back during a final weapons drill, surveying the group. Sweat-coated, exhausted, and grinning with accomplishment, they were ready—or as ready as they could be—when the Ranger company finally arrived.

Jack allowed himself a small, private smile. For the first time in weeks, he felt a flicker of satisfaction. They had come a long way together, and the bond they had forged would carry them through whatever came next.

The base buzzed with new energy the day the Ranger company arrived. Trucks rolled through the gates, boots hit the gravel, and voices carried across the compound as seasoned soldiers filed in. Jack stood with Davis, Hill, Kim, and Reilly, watching the organized chaos settle into order.

It wasn't long before a runner came up and barked, "Staff Sergeant Mercer, Captain Walker wants you in his office. Now."

Jack straightened, gave his men a nod, and followed.

Inside, the company commander, Captain James Walker, sat behind a desk piled with folders and dispatches. His uniform was pressed but already rumpled by the day's work, and his sharp eyes studied Jack as soon as he entered.

"Staff Sergeant Mercer," CPT Walker began, motioning him forward. "I've been getting reports. Seems you've been keeping yourself and a handful of men busy while we were on the water."

"Yes, sir," Jack replied. "Didn't see any sense in letting the time go to waste."

CPT Walker leaned back, folding his hands. "I like that. Rangers don't sit on their hands. Our mission here isn't going to be straightforward fighting. We'll be tasked with reconnaissance, long-range patrols, and deep insertions behind enemy lines. It's going to take initiative, endurance, and discipline."

He let that sink in for a moment before' continuing. "I've seen what you've done with Davis, Kim, Reilly, and Hill. Turning a green bunch into something sharp in just a few weeks isn't easy. That tells me you're the kind of NCO who knows how to shape men."

Jack stood quietly, unsure where the captain was going, but attentive.

CPT Walker leaned forward now, his tone firm. "So, here's what I'm going to do. I want that group of yours to remain together as its own fire team within the company. You'll answer directly to your platoon leader, of course, but I want your team to operate as a specialized element. You've got chemistry, and I'd be a fool to break that up."

Jack gave a curt nod. "Understood, sir. I'll make sure they're ready for whatever comes."

CPT Walker's gaze hardened, though there was respect behind it. "Good. Because what's coming won't be easy. Korea's about to get a whole lot worse, and I'm going to need men I can rely on to push further than the rest. You make damn sure your team is one of them."

"Yes, sir," Jack replied, voice steady.

As he turned to leave, CPT Walker added, "Oh, and Mercer… I don't know what your story is, but I've seen men like you before. Quiet, steady, dangerous when they need to be. Don't prove me wrong."

Jack left the office, the weight of the new responsibility settling in his chest. Outside, he saw his team waiting, their eyes curious. For the first time, Jack allowed himself the faintest of smiles.

Jack stepped out into the sunlight, squinting as he spotted Davis, Hill, Kim, and Reilly leaning against a wall nearby. They were pretending to look casual, but their posture told him they'd been waiting on him the whole time.

Davis pushed off the wall first. "So, Sarge—what's the word? We in trouble already?"

Jack shook his head, lips quirking into the faintest grin. "Not hardly. We've been promoted."

The four exchanged quick, puzzled looks. Reilly scratched the back of his neck. "Promoted? How's that work? We didn't do nothing' yet."

Jack folded his arms, letting the suspense hang for a moment. "CPT Walker's seen what we've been doing—training, shooting, pushing ourselves. He's decided we're going to operate as our own fire team inside the company. That means we stick together, and we'll likely be sent on missions the others won't."

Hill let out a low whistle. "So, we're the guinea pigs."

Kim grinned, elbowing him. "Nah, man—we're the tip of the spear."

Davis, steady as always, nodded thoughtfully. "Sounds like the captain's got plans for us. That means we don't get to screw up. Not once."

Jack's expression sobered. "He's right. We're Rangers. That means we'll go further, hit harder, and take risks the regular infantry won't. Some of those risks are going to be bad. Real bad. But we've got each other—and I'll make damn sure we're ready."

The men fell quiet, the weight of his words settling in. Finally, Reilly broke the silence with a crooked grin. "Guess that means more of your crazy PT, huh?"

Jack allowed himself to chuckle. "You guessed right. So, eat up tonight, boys. Tomorrow morning, we start earning that promotion."

The team groaned, but beneath it, Jack could see the pride in their eyes. They weren't just soldiers anymore. They were something more—and they knew it.

The men were still grumbling about "crazy PT" when Jack clapped his hands together.

"Don't just stand there," he said. "Grab your gear—we're not waiting until morning."

Davis blinked. "Tonight? We just got told we're a team, and you're already trying to run us into the ground?"

Jack's grin was quick and sharp. "Exactly. The enemy's not going to wait until morning. Neither do we."

Hill groaned, Kim muttered something in Korean that made Reilly laugh, but all four of them pushed off the wall and followed Jack.

They stashed their personal gear in the billets, cinched down boots, and stepped out under the dusky sky. Jack led them out past the edge of camp to a stretch of uneven ground where the tree line began.

"All right," he said, standing before them like an old hand on parade. "Tonight's light. Three miles, full kit. Tomorrow, we will shoot until you can strip and clean your rifles blindfolded. End of the week, you'll be running circles around the rest of this company."

Reilly looked at the others, smirk tugging at his mouth. "We're not Rangers yet, boys—we're Mercer's Misfits."

The name stuck, the rest of them laughing as they started jogging into the gathering dark, their boots thudding in unison.

Jack ran to the front, breathing steadily, pushing them harder than they thought possible. For him, it wasn't just training, it was a way to keep his mind from drifting back to Patricia, to the way her hand lingered in his when they'd said goodbye. The ache in his chest eased, just a little, each time he heard the men behind him push through their exhaustion instead of quitting.

By the time they circled back to camp, sweat-soaked and gasping, Jack could see it in their eyes: pride, grit, and a growing bond. They weren't just five men anymore. They were a team.

And Jack knew they'd need that bond soon enough.

Over the next few weeks, Mercer's little team trained harder than anyone else on base. From sunup to long past sundown, they ran the ranges until their shoulders ached, drilled patrolling techniques through the thick brush, and pushed themselves with punishing PT sessions Jack seemed to conjure out of thin air. Sweat, blisters, and bruises became routine, but so did the bond between them. Davis, Hill, Kim, and Reilly began moving like one, their confidence in Jack's leadership growing with every passing day.

Word of their relentless drive spread quickly. Men in the other platoons started calling them *Mercer's Misfits*—half in jest, half in respect.

One morning, under the hard, rising sun, the company gathered for a briefing. Dust hung in the air as squads and platoons fell in, leaders shouting over the clatter of weapons and shifting packs.

CPT Walker, the Ranger company commander, strode forward, his eyes sharp as they swept over the formation before settling on Jack and the men at his side.

"Staff Sergeant Mercer," he barked, voice carrying over the restless chatter.

Jack stepped forward, boots snapping together. "Sir."

CPT Walker folded his arms. "I've been hearing about what you and your men have been up to. Extra PT, and Extra Range time. Pushing yourselves harder than anyone else. Some of the boys are saying you think you're special." His mouth twitched with the faintest grin. "So, let's find out."

He let the silence stretch a moment before continuing. "Tonight, during field maneuvers, you and your men will attempt to infiltrate First Platoon's perimeter. And not just slip through it—I want you to snatch either their platoon leader or their platoon sergeant and drag him back here without being caught."

The formation erupted into murmurs and chuckles.

CPT Walker's grin sharpened. "Do it, and you'll earn this company's respect. Fail, and you'll just be another pack of loudmouths."

Jack didn't hesitate. "Yes, sir. We'll get it done."

Behind him, Davis smirked at the challenge, Hill muttered a curse, Kim cracked his knuckles, and Reilly shook his head, half amused, half worried.

Jack turned slightly, voice low but firm. "Gear up, Misfits. Tonight, we show them who we are."

That night the forest around the training grounds was alive with low murmurs, shifting boots, and the faint glow of smoldering cigarettes. First Platoon had been warned. CPT Walker wanted to give Mercer's Misfits the hardest test possible, so he'd tipped off the platoon leadership. They had doubled their sentries, tripled their wire, and set up noise traps with tin cans strung on trip lines.

Mercer's squad crouched in the tree line, blackened faces hidden under the moonlight, watching the platoon's perimeter with patient eyes. Jack scanned the defenses, seeing the patterns, the weak points others might miss. Davis leaned close, whispering, "They're waiting for us. Whole damn platoon is on edge."

Jack's voice was calm, almost casual. "Good. That means they'll be sloppy. Overconfidence always leaves cracks."

Hill muttered, "Hell of a lot of cracks to find."

Reilly swallowed hard, but Kim just gave a silent nod, his sharp eyes tracking every movement of the distant sentries.

Jack smiled faintly. "Follow my lead. Remember—quiet is life."

They began their crawl. Slow, deliberate, inch by inch. The night was alive with cicadas and rustling leaves, masking the faint scrape of cloth and metal. Twenty yards from the wire, Jack froze them with a hand signal. Ahead, a sentry leaned against a tree, rifle slung lazy, muttering to another man in the shadows. They thought they were alert, but Jack saw it—the half-second lag between glances, the complacency that comes from thinking you're ready.

Davis whispered, "How are we getting past them?"

Jack smirked. "We're not. They are."

He tossed a pebble far to the right. The faint crack drew both heads instantly. "What the—?" The sentries shifted, rifles raised toward the sound. In that heartbeat of distraction, Jack waved the team forward. They slipped past, pressed to the ground, the darkness swallowing them whole.

Minutes stretched like hours as they wove through the perimeter, avoiding tripwires, using the shadows of trees and tents. Twice, Hill nearly tripped a can-line, only saved by Jack's hand clamping his shoulder.

At last, they reached the heart of the camp. The platoon command post sat under a camouflaged tarp, a dim lantern casting light on a folding table where the Platoon Leader and Platoon Sergeant sat, maps spread before them. A runner had just left, probably checking the lines.

Jack whispered, "On me."

The Misfits moved quietly. Davis slipped left, Kim right, Reilly and Hill trailing Jack. The flap rustled just slightly as Jack entered, pistol rose but safety on. "Evening, gentlemen."

The lieutenant and sergeant froze. "What the—"

Before they could shout, Davis and Kim rushed in, clamping hands over mouths, pinning arms. Jack leaned in close, low and confident voice. "CPT Walker sends his regards."

Minutes later, First Platoon's entire perimeter erupted in chaos as the Misfits strode back toward CPT Walker's CP, the lieutenant and platoon sergeant gagged and tied, struggling but unharmed. Sentries shouted, whistles blew, boots thundered, but it was too late—their leaders were gone.

When they emerged into the glow of lanterns at the Ranger company headquarters, CPT Walker stood waiting, arms folded, a wide grin splitting his face. The rest of the company gathered fast, men craning necks to see. Laughter, shouts, and even a few cheers broke out as the Misfits dragged in their prizes.

CPT Walker raised his hand for silence. "Well, I told them you were coming. I gave them every chance to stop you. And yet here you are, Staff Sergeant Mercer… with both their platoon leader *and* their platoon sergeant."

He stepped closer, his eyes sharp with approval. "Looks like your little team just became this company's special operations section."

The men erupted in cheers again, and Jack glanced at his squad—Davis smirking with pride, Kim standing stone-still but pleased, Hill grinning ear to ear, Reilly shaking his head in disbelief. Jack let a rare smile slip.

"Good work, Misfits," he said quietly. "This is only the beginning."

Chapter 10: Bound in Silence

In mid-December, CPT Walker called the company together in the chill of the Japanese winter. The men gathered in formation, their breath hanging in the cold air, silence falling as their commander stepped forward.

"Listen up," CPT Walker began, his voice carrying with a weight that settled in every chest. "Orders just came down from Division. On the twenty-third of December, this company will deploy to Korea. Once there, we'll be attached to the 2nd Infantry Division."

A ripple moved through the ranks—excitement, nerves, determination. They all knew this was coming, but hearing the date spoken aloud made it real. Korea wasn't just a place on the map anymore. It was waiting for them.

Jack stood among the men, his eyes narrowing slightly. Another war. Another fight. For him, for the Misfits, there was no turning back now.

After CPT Walker dismissed the company, the men broke into knots of conversation. Some were joking to hide their nerves, others already making mental lists of gear they wanted squared away. Jack turned to his small circle—his Misfits.

Sergeant Frank Davis shook his head, letting out a low whistle. "Christmas in Korea. Hell of a gift from Uncle Sam." He tried to keep it light, but there was heaviness in his eyes that betrayed him.

PFC Tom Hill, tall and broad-shouldered, shifted his weight uneasily. "I knew it was coming, but… damn. It feels different when you hear the date. My old man always said December's a cursed month for wars."

Reilly, the youngest of the bunch with a grin that always came too easy, finally broke the silence. "Guess this'll be one hell of a story someday, eh? Survive Korea, I'll never pay for a drink again." But the false bravado didn't fool Jack.

Kim stood a little apart, quiet at first, then spoke in a steady tone. "My parents still have family in the South. I don't know if they're safe. But if we're going… then maybe I'll find out." His eyes flicked away, emotion hidden behind discipline.

The others fell silent, sobered by his words.

Jack let them speak, let the worries, the bravado, and the doubts settle. Then he squared his shoulders and looked each of them in the eye.

"We've trained harder than anyone else in this company. You've put in the work, and you're ready. Doesn't matter if it's Korea, France, or the damn moon—we'll handle it. Together."

The men straightened, their nerves settling into something stronger. Davis gave a curt nod, Hill exhaled slowly, Reilly managed a grin that looked a little less forced, and Kim clasped his helmet under his arm, jaw tight but resolute.

Jack knew the fear would come and go, as it always did. But he also knew his Misfits would follow him into the fire—and he'd make damn sure they all came out on the other side.

Over the next couple of days, Mercer's Misfits moved like a well-oiled machine. Gear was checked and double-checked, weapons cleaned and re-cleaned, packs packed to the ounce. The sound of straps tightening, metal clanging against wood, and low murmurs of coordination filled the air as they prepared for the move to Korea.

Jack moved among them, giving pointers, adjusting straps, tightening knots, and silently assessing each man. Then he called them together.

"All right, Misfits," he said, holding up a pack. "Put it on. Now, everyone jump up and down."

The men complied. Boots thudded against the floor, straps rattled, and metal buckles clanged loudly. Jack raised an eyebrow. "Exactly what you don't want when sneaking behind enemy lines. Now, your task: as a team, come up with a Standard Operating Procedure so that when you put this pack on and jump, it makes no noise. Every strap, buckle, and flap silent. And I want it done before lights out tonight."

Davis shook his head, laughing. "You're insane, Jack."

Hill grinned, rubbing his shoulder. "I like it. Makes sense though."

Kim's sharp eyes scanned the pack, noting points of friction and possible slack. Reilly muttered something about duct tape and padding.

Jack smiled faintly, watching them start to work together, testing the packs, adjusting straps, wrapping cloths over buckles, and quietly critiquing one another. "This," Jack thought, "is how we stay alive in Korea."

Even as he focused on their readiness, his thoughts drifted back to Patricia. The memory of her laughter, the warmth of her hand in his, and the unspoken bond they shared pressed on him like a weight. He knew he needed to see her again before leaving—but not in the hospital. Too many eyes, too many questions. He wanted somewhere private, somewhere he could talk to her freely, even if just for a few hours.

Jack made a mental note to find that place. He couldn't leave without it, couldn't leave without seeing her once more—even if only for a moment that would have to last him through the battles ahead.

That evening, after the Misfits had finished their SOP drill and quieted down, Jack slipped away to a quieter corner of the base. He needed a plan—a place to meet Patricia where no one would watch, no one would question.

He thought of the small Japanese garden near the hospital grounds, a place often overlooked, tucked behind tall hedges and lanterns. It was serene, private, and far enough from the bustle of patients and personnel that they could talk without interruption.

Jack pulled a sheet of paper from his pocket and quickly scribbled a note:

"Patricia, I need to see you—somewhere private. Meet me in the garden behind the hospital at 1900. Alone. —Jack"

He folded it neatly, slipped it into an envelope, and handed it to a trusted orderly who promised to deliver it discreetly.

Jack stood for a moment, staring at the moonlit grounds, imagining her reading it, imagining her understanding of the urgency without any explanation. He couldn't tell her about the Misfits' deployment yet, couldn't reveal his secret or the full scope of what lay ahead—but he could have this one moment with her.

As he returned to his quarters, a quiet resolve settled over him. No matter what happens in Korea, he would see her tonight. And for a few precious hours, nothing else would exist.

At exactly 1900, Jack made his way to the small Japanese garden behind the hospital. The soft rustle of leaves and the faint glow of lanterns created a secluded world far removed from the hum of the base.

Patricia was already there, standing near a stone bench, her hands clasped in front of her. When she saw him, her lips curved into a small, relieved smile.

"Jack," she whispered as he approached, "you made it."

He closed the distance between them and took her hands gently. "I had to. Before… everything changes."

For a long moment, they just looked at each other, the unspoken bond between them filling the quiet night. Jack traced a thumb over her knuckles, feeling the warmth of her skin.

"I wanted to see you," he admitted, voice low. "Not at the hospital, not around anyone else. Just… us."

Patricia nodded, her eyes glistening. "I understand. I wanted the same."

Jack drew her closer, and their embrace deepened. The tension of weeks apart melted as they held each other. Then, almost instinctively, their lips met, a soft, lingering kiss that spoke of longing, fear, and hope all at once.

Pulling back slightly, Jack rested his forehead against hers. "I don't know what the next weeks will bring, Patricia. But I promise you—I'll come back."

She pressed her hand to his chest, feeling the steady beat of his heart. "I'll wait for you, Jack. No matter what."

For a few precious minutes, they existed only for each other, the world outside forgotten. And though Jack knew the coming days would demand everything from him, this moment gave him strength, courage, and a reminder of what he was fighting for.

Finally, with a reluctant sigh, they separated. "I have to get back," Jack said softly. "But… thank you. For this. For us."

Patricia smiled, brushing a stray lock of hair from his forehead. "Be careful, Jack. And come back to me."

Jack gave a last lingering look, then disappeared into the shadows, leaving the garden empty but still echoing with the warmth of their connection.

Jack slipped back into the barracks under the dim glow of lanterns, careful not to draw attention. The Misfits were still at work, double-checking gear, packing weapons, and reviewing maps. The soft clatter of metal and the low murmur of coordination filled the room.

Davis looked up as Jack entered, a questioning eyebrow raised. "Everything okay, SSG?"

Jack nodded, sliding a pack onto the table. "Fine. We move out in two days. I want nothing left to chance."

Hill grinned, trying to lighten the mood. "Sounds like someone had a busy night."

Jack smirked faintly but didn't answer. He moved through the group, checking straps, tightening knots, and offering pointers on how to balance weight and silence the gear. "Remember, once we're in Korea, everything counts. Quiet movement, clean weapons, and trust in each other. That's how we survive."

Reilly adjusted his pack nervously. "Don't worry, SSG. We've got your back."

Kim nodded silently, methodically inspecting his rifle, his focus unbroken.

Jack surveyed them all, pride mixing with the weight of responsibility. They were ready—physically and mentally—but he knew the coming battles would test more than their strength. Still, his thoughts drifted, unbidden, to Patricia. The warmth of her smile, the pull he couldn't deny.

He knew then what he had to do. Before they shipped out, he would see her one more time. He would tell her the truth—the secret he had carried alone since the war. If he was going to Korea, he couldn't leave without her knowing who he really was.

"Finish your packs," he instructed, voice firm. "We've got two days to tighten this up. Rest when you can but stay sharp. This is the start of something big, Misfits. Make sure we're ready for it."

As the men went back to work, Jack steadied his nerves. Soon, he'd be facing combat again. But first, he had one more battle of the heart.

The next day was a blur of routine, the kind of tasks that felt oddly mundane against the weight of the upcoming deployment. The Misfits moved with precision, checking gear, finalizing weapons, and coordinating last-minute details. Meanwhile, Jack spent time signing paperwork—life insurance forms, wills, and other official documents that seemed to shrink him to the reality of mortality.

As he scribbled his name on the forms, a thought took root. If anything happened in Korea, he wanted Patricia to have everything. His mind raced with memories of her, the warmth of her presence, and the bond they shared.

By late afternoon, Jack had made his decision. He would leave everything—his belongings, his small savings, even the Mercer farm if needed—to Patricia. That evening, he would give her the paperwork personally, not just as a formality, but as a gesture of trust and love.

And more than that, he would share the truth about his past—the secret he had carried for years. Everything about the experiment, about his altered aging, about the life he had lived in the shadows of history. She deserved to know, and he could no longer keep it from her.

As he carefully organized the documents into a small leather folder, Jack's mind lingered on Patricia's face. He imagined her reading the papers, then looking up at him, and he felt a mix of anticipation and fear. The truth could change everything—but if it was going to, he wanted her to know him completely before he left for Korea.

Jack tucked the folder under his arm. Two things were clear: the Misfits were ready for the mission, and he had to see Patricia tonight. No more delays.

That evening, as the sun dipped below the horizon, Jack made his way to the small Japanese garden behind the hospital. Lanterns glowed softly, their warm light spilling across the stone path. He carried the leather folder carefully, each step heavier with the truth he was finally ready to share.

Patricia was waiting by the bench, her hands clasped in front of her. When she saw him, her face softened. "Jack… what's wrong? You sounded urgent in your note."

He took her hands gently, pressing the folder into them. "This is my will, my insurance paperwork. If something happens in Korea… it all goes to you."

Her brows drew together. "Jack…"

"Wait," he said, his voice steady but low. "There's more. You deserve the truth about me. During the war, the Germans did something to me. An experiment. It changed me—I age differently. Slower. One year for every ten. Sam and Doc are the only two people who know. Since the war ended, I've drifted, never staying in one place more than a few months because… well, how do you explain not aging?"

Patricia's breath caught, but her eyes didn't turn away. Instead, they softened, almost as if she'd been waiting for this moment.

"I need you to know something too," she whispered. "The same thing happened to me. Years before I escaped Germany. I was just a teenager when they experimented on me. That's why… I don't age like others. I've kept it hidden all these years. I never thought I'd meet someone like me."

Jack's grip on her hands tightened, a tremor of relief and awe coursing through him. "Patricia… all this time I thought I was alone."

Her eyes shimmered in the lamplight. "So did I."

For a long moment, they stood in silence, holding each other's gaze. The world seemed to fade, leaving only the two of them two souls bound by fate and circumstance.

Finally, Jack spoke, his voice hushed but firm. "I can't leave without asking you this. Patricia… marry me."

Her lips parted, her breath catching in her throat. "Jack…"

"We'll have to keep it secret," he said quickly, urgency and hope in his eyes. ["You're an officer, I'm enlisted. They'd never allow it. But none of that matters to me. I don't care about rules or regulations—I care about you. If we keep it between us, we can make it work. Just say yes."]

For a heartbeat, she only stared at him, her chest rising and falling as emotion welled up inside her. Then, slowly, a smile broke through the tears forming in her eyes.

"Yes," she whispered. "Yes, Jack. I'll marry you."

He pulled her into his arms, their lips meeting in a kiss that was both fierce and tender, sealing their promise under the quiet glow of the garden lanterns. For the first time in years, Jack felt whole.

Their kiss lingered long enough that Jack felt his chest ache with the weight of both love and urgency. When they finally parted, Patricia reached up, brushing her fingers against his cheek.

"I don't want you going into Korea with nothing of me," she whispered. She unclipped the small silver cross that hung from her necklace a simple piece, worn but cherished. Pressing it into his palm, she closed his fingers around it. "This has been with me since I left Germany. It kept me strong when I thought I wouldn't make it. Now… I want you to carry it."

Jack looked down at the cross, his throat tightening. "Patricia… I can't take this."

"You can," she said firmly. "Because I'll still feel it with me, even if it's not around my neck. And when you look at it… you'll know I'm waiting for you."

For a long moment, Jack couldn't speak. He pulled her close again, resting his forehead against hers. "Then you'll need something of mine," he said, his voice rough. Reaching into his duffle, he pulled out a small, worn patch—his old paratrooper jump wings, the cloth frayed but intact. "These have been with me since Normandy. I survived that day because of men like Rizzo… because of the bond we shared. Now, I want you to have them."

Patricia cradled the patch carefully, tears glistening in her eyes. "I'll guard them with my life."

Jack kissed her once more, slow and deliberate. "When I come back, we'll make this official. No more secrets. No more hiding."

Patricia nodded, clutching the patch to her chest as though it was a promise carved in stone. "I'll hold you to that, Jack Mercer."

The garden around them was silent save for the faint rustle of the winter wind through the trees, carrying with it the weight of a vow that neither would ever forget.

Jack walked Patricia back to her quarters, the night quiet around them except for the faint sound of boots on gravel in the distance. As Jack held Patricia close, their whispered promises bound them together in a way neither could explain but both felt deep in their bones. For a fleeting moment, the war, the uniforms, and the secrecy all faded away, leaving only two souls who had finally found someone who understood.

As he made his way back to camp, the cool December wind cut across the streets of the base, sharp with the scent of the sea. Jack tucked his collar higher, his thoughts racing between Patricia's face and the reality of what lay ahead. Tomorrow, the Ranger company will be gone, bound for Korea and war.

He whispered to himself, almost like a vow, "I'll come back to you."

With that, Jack squared his shoulders and stepped back through the gates, the cross against his heart reminding him that for the first time in years, he had something worth fighting to return to.

Part 3: Fire and Frost

Chapter 11: Into The Cold

The harbor buzzed with activity as the Rangers boarded the transport ship, crates of ammunition and rations stacked high on deck. SSG Mercer led his men aboard, checking straps, adjusting packs, and securing his M1A1 carbine. His kit was a curious mix of old and new—worn WWII gear alongside experimental prototypes. The folding-stock paratrooper M1A1, the battered jump knife from Normandy, a reinforced harness designed to redistribute weight, lightweight alloy canteen, and field boots with experimental soles—everything combined into a loadout that drew both admiration and amusement.

As the ship pulled away, the men started noticing. "SSG Mercer," Reilly said, shaking his head, "I swear half your kit belongs in a museum."

Hill laughed, slapping the side of SSG Mercer's harness. "And the other half looks like it was delivered by a scientist straight out of the future. How do you even move in that thing?"

Kim smirked quietly. "I'll give it to you, SSG Mercer. You look ready for anything… even if nobody else knows what half of it does."

Davis shook his head, grinning but not missing a beat as he checked the squad's gear. "Only SSG Mercer would mix Normandy with next-generation prototypes and somehow make it look normal. Don't ask me how, but it works for him."

Jack allowed a faint smile. "Old gear never failed me. New stuff might just save my life. You stick around long enough, you'll see."

The others laughed, but it was good-natured ribbing. They respected him too much to challenge the wisdom in his choices.

Below deck, the men stowed their remaining gear and found whatever space they could for rest. The ship rolled gently in the waves, the sound of water against the hull mingling with creaking timber and the low murmur of conversation. Jack leaned back against a bulkhead, M1A1 across his lap, the prototype harness shifting slightly as he settled in.

His mind drifted to Normandy, the jump, Rizzo and the men lost in the field, and the chaos of that day. He pressed his hand to the silver cross in his pocket, Patricia's gift grounding him in a way nothing else could.

The voyage stretched on, hours measured in the hum of engines and the rhythm of waves. His team continued giving him a hard time with the mixed gear, teasing and joking, but beneath the laughter was admiration. SSG Mercer was different, yes—but they knew it wasn't eccentricity that made him stand out. It was an experience.

The sun dipped low, casting long shadows across the deck. Jack adjusted his harness once more, checked the straps on his prototype canteen, and exhaled. Korea awaited at the end of this journey. And whatever came, the cross in his pocket reminded him there was someone worth returning to, someone waiting far beyond the horizon. Below deck, the Rangers had claimed small corners to rest, stow gear, or simply catch a moment of quiet. Jack leaned against a bulkhead, adjusting the straps on his prototype harness while Davis and Hill sprawled nearby, tossing small objects back and forth to pass the time. Reilly fiddled with the M1A1 magazine, and Kim quietly studied the layout of the ship, eyes scanning the bulkheads and railings as if memorizing every inch.

"SSG Mercer," Reilly said, leaning over and tapping the harness, "seriously, that thing looks like it was designed by an Alien. You planning to fly us to Korea yourself?"

Hill laughed, slapping the side of Mercer's duffle. "Yeah, and don't forget the jump knife from Normandy. Gonna need that for the dragons over there, right?"

Jack shook his head with a small smile. "Keep it up, and I'll make you carry it all next time."

Davis chuckled, eyes never leaving the squad as he spoke. "You know, SSG Mercer, I've seen a lot of soldiers with fancy kit, but nobody pulls this off the way you do. Part museum, part lab experiment—and somehow it works. I don't know how, but it does."

Kim finally spoke, his tone quiet but firm. "I'll admit, SSG Mercer… the harness alone might save our backs on the march. Maybe all of it will."

The ribbing eased a little as the others laughed. For all their joking, they respected him—and they knew it. His odd mix of WWII gear and prototype equipment wasn't vanity or eccentricity. It was experience forged in the fires of combat, lessons learned the hard way, and survival instincts built over decades.

Hours stretched on as the ship cut through gray waves, the engine's hum a constant reminder that Korea waited at the end of the voyage. The camaraderie grew with each shared joke, each teasing comment about Mercer's "time-traveling kit," and the quiet moments of watching the ocean pass beneath the hull.

At one point, Reilly nudged Hill and whispered, "I'll bet SSG Mercer could teach half these kids back home more than the Army ever could."

Hill nodded. "Yeah… he's not just leading a team. He's building us into something we haven't even realized we can be yet."

Jack, hearing the quiet admiration, allowed himself a faint smile. He pressed a hand against the silver cross in his pocket, thinking of Patricia and the promise he had made. Whatever challenges awaited them in Korea, he would meet them with this team at his side—and a memory of someone worth returning to waiting far beyond the horizon.

The sun dipped lower, painting the deck in long, cold shadows. The ocean stretched endlessly ahead, but soon land—and the reality of war— would break the horizon. The Rangers settled into a rhythm for the remainder of the voyage, joking, checking gear, and quietly steeling themselves for what was to come.

The gray morning air was heavy with mist as the transport ship crept into Pusan Harbor. Jack, standing near the railing, adjusted his mixed WWII and prototype gear one last time. The wind tugged at his coat and rattled the straps on his pack, but he barely noticed. His eyes scanned the coastline, the city in the distance, and the hills beyond, knowing that what awaited them was far from welcoming.

"SSG Mercer," Reilly said from beside him, squinting through the mist, "I don't like the look of that shore. Looks like trouble waiting to greet us."

Hill nudged him with an elbow. "Trouble? You think it's worse than Normandy beach? Come on, it's just a harbor."

Jack allowed himself a faint smirk. "You'll find out soon enough. Stay sharp. This isn't a training exercise anymore."

The rest of the Rangers were busy double-checking packs, rifles, and harnesses, their movements tense but precise. Davis walked past, giving SSG Mercer a brief nod. "Everyone's counting on you, Sarge. Make sure we all make it off this boat in one piece."

The gangway clattered as it was lowered, and the Rangers began moving toward it. The smell of salt, diesel, and war hung thick in the air, mixing with the distant echoes of artillery and small arms fire from the front lines' further inland.

SSG Mercer led his team down the ramp, boots striking the metal as the ship rolled slightly under their weight. The harbor workers and military personnel shouted orders over the din, the chaos amplifying every sound. Reilly nearly stumbled, but Mercer's hand on his shoulder steadied him.

"Thanks, SSG Mercer," Reilly muttered under his breath.

Jack only nodded. "Keep moving. Eyes open."

The dock was crowded, trucks and jeeps lined up to transport troops and gear, and the occasional flare of smoke from a distant skirmish painted the hills in a ghostly orange. The tension in the air was palpable. For the first time in months, the training, the jokes, and the camaraderie felt small against the enormity of what lay ahead.

As they moved inland, Jack couldn't shake the thought of Patricia. He felt the weight of the silver cross in his pocket, the promise he had made to her, and a pang of longing. He couldn't know whether or if he would see her again. The secret they shared, and the lives they led, kept them apart even as he carried her with him in every step.

Davis, noticing Jack's momentary distraction, clapped him on the shoulder. "Focus, SSG Mercer. We'll deal with everything else once we're settled. Right now, it's boots and eyes open."

Jack took a deep breath, letting the harbor noises, the distant gunfire, and the smell of salt and diesel ground him. He tightened his harness, adjusted the prototype straps one more time, and led his men forward. Korea awaited.

And the war, as brutal and unforgiving as any he had known, had already begun.

The convoy rumbled along the narrow roads leading inland from Pusan, trucks kicking up dust and mud as the Rangers sat tightly packed in the beds, rifles across their laps. Jack scanned the passing countryside, noting every rise, every tree line, every potential ambush point. The hills were low and brown, winter stripped them bare, but he knew that hidden in those folds could be enemy forces.

"SSG Mercer," Reilly said, voice low, "you think the North Koreans are waiting for us out there?"

Jack glanced at him, expression unreadable. "Probably. But we don't know where. That's why we watch, we move, and we stick together."

Hill nudged Reilly with a grin. "Don't worry. If anyone's going to keep us alive, it's sarge and his museum-of-the-future gear."

Reilly laughed softly, shaking his head. "Yeah, yeah, I get it. We're in good hands."

Kim sat silently, eyes fixed on the passing terrain, but even he couldn't hide the faint nod of respect. Davis, ever methodical, checked the straps on the prototype harnesses the men were wearing, muttering to himself about load distribution and noise reduction.

Jack allowed himself a quiet smile at the banter. The teasing about his WWII and prototype gear eased some of the tension, but he knew the real test was coming. Korea was not a training field.

As the convoy approached Kijang, they passed farmers working frozen fields, curious civilians peering at the unfamiliar trucks and uniforms. Some waved tentatively; others hurried indoors. The Rangers kept their eyes on the roads and ridges, wary of ambush or sniper fire.

The base at Kijang came into view—a modest collection of barracks and tents surrounded by a perimeter of sandbags and makeshift defensive positions. Smoke from cooking fires drifted lazily into the sky, and a handful of soldiers moved about, issuing orders and preparing for the arrival of the new unit.

"SSG Mercer, the command tent is over there, and your squad tent is to the right of the command tent," a young lieutenant called, clipboard in hand. "Welcome to Kijang. Things are tense, but you'll get your bearings quickly."

Jack nodded, already thinking ahead. "Copy that. Keep your men ready and moving. Eyes open, ears open."

The convoy slowed, and the Rangers disembarked, hauling packs and checking weapons as they moved toward the tents. Jack noticed the nervous energy in the younger men—this was their first time in a real war zone, and it showed.

"SSG Mercer," Reilly said quietly as they walked, "how do you stay so calm? I mean… this is Korea, not some exercise back at Fort Ord."

Jack glanced at him, letting the weight of his experience show just enough to impress, but not intimidate. "Experience teaches you two things, Reilly. One, fear is normal. Two, you don't let it control you. Keep your head, follow your training, and watch your team. Everything else comes after."

As they reached the squad tent, Jack felt the familiar pull of responsibility settle over him. His team, his gear, and his years of experience all of it would be tested here. Korea was a different battlefield, but the principles were the same: watch, move, survive.

And for the first time since leaving Japan, Jack allowed himself a moment to think about Patricia again, tucked safely in his pocket and in his heart, a reminder of everything worth fighting for.

The Rangers hauled their packs across the frozen, mud-choked ground of Kijang, finally reaching the small, crowded compound. The company had been attached to the 23rd Infantry Regiment, a move meant to consolidate forces, but it had created immediate friction. Food and water were in short supply, and tempers ran high.

"Welcome to real war, boys," muttered SSG Carver of the 23rd Infantry, leaning against a sandbag barricade, eyes narrowing as he sized up Jack and his team. "Looks like a fresh batch of Rangers wandered into my backyard. But you've never fired a real rifle in anger before."

Jack's jaw tightened, but he stayed calm. "We've seen combat," he said evenly, letting the weight of his rank and demeanor speak for itself.

Carver laughed, a harsh, rasping sound. "Oh, I don't doubt that you *say* that. But I've been in the thick of it, Mercer. Normandy, the Bulge, Korea… you're green until proven otherwise. And from what I hear, your fancy little 'team' has more gadgets than sense."

Reilly muttered under his breath, fists tightening, but Jack held up a hand. "Carver," he said, voice calm but firm, "I don't need to prove anything to you. My men and I will do our jobs. You focus on yours."

Carver's eyes narrowed, and he spat on the ground. "We'll see, Mercer. Don't think the brass assigned you here so you could show up and be a hero. I've earned my stripes the hard way. You just showed up with a neat uniform and some prototype toys."

Davis stepped forward, adjusting his pack strap. "SSG Carver, we respect your experience, but we're all here for the same reason—to survive, to do our jobs, and to watch each other's backs. That includes your unit."

Carver's gaze flicked between Davis and Jack, lingering on the Ranger's calm, commanding presence. He sneered, but there was a hint of hesitation. He knew Mercer wasn't just talking.

Jack exhaled slowly. "Carver," he said, voice low but cutting, "you want to test us, fine. But don't mistake caution for weakness. You'll see what we're capable of soon enough. Until then, save your judgment and keep your men in line. We'll handle ours."

The 23rd Infantry SSG grunted, stepping back, clearly unwilling to escalate further under Mercer's calm authority. The tension lingered, thick as the fog rolling across the compound, but Jack knew he had made his point. Respect in this place wasn't given it was earned. And he had already survived enough to know how to earn it quickly.

As the Rangers settled into their crowded barracks, the teasing and camaraderie among his own men returned. Reilly nudged him. "SSG Mercer, you handled that guy like a pro. Didn't even break a sweat."

Jack allowed himself a faint smile, adjusting the straps on his prototype harness. "It's not about breaking a sweat. It's about keeping the team alive.

Remember that boys. Watch each other's backs, and the rest will follow."

Outside, the wind howled across the perimeter, carrying the faint scent of distant smoke and the promise of a long, bitter fight to come.

Three days on Korean soil had been enough to wear down even the toughest of men. The Rangers had moved gear, scouted the immediate perimeter, and adjusted to the relentless cold, mud, and scarcity of supplies. Tension with the 23rd Infantry remained, but SSG Mercer's calm leadership and experience kept his team focused and prepared.

The morning was gray and overcast, the wind carrying the faint smoke of distant skirmishes across the compound. Jack finished checking the straps on his harness one last time, ensuring the M1A1 was secure, and the prototype gear properly fastened.

"SSG Mercer," a runner called, appearing from between the tents. "CPT Walker wants you in the command center. Now."

Jack nodded, his expression unreadable. "Copy that. Let the men know. Keep your eyes open."

As he walked toward the command center, the Rangers watched him go, knowing that when

Mercer left, he carried not just his own experience but the safety and readiness of the entire team.

The compound faded behind him, replaced by the dirt roads leading to the command tent. Jack's mind shifted to the unknown—missions, reconnaissance, and the brutal reality awaiting them beyond the perimeter. He adjusted the prototype harness one final time, tightened the strap on his M1A1, and pushed forward.

Inside CPT Walker's command center, maps were spread across the table, radios hummed, and officers moved with purpose. Jack's presence drew quick glances and brief nods of acknowledgment. He knew what was coming: a mission that would test everything he had taught his men and all that he had learned himself.

Jack exhaled slowly, feeling the familiar weight of responsibility, settle onto his shoulders. The fight was just beginning.

Chapter 12: Through the Wire

CPT Walker's command center was dimly lit, the scent of tobacco and cold metal hanging in the air. Maps of the surrounding terrain were spread across the table, pins marking known enemy positions and lines of advance. Jack stepped inside, boots clicking against the wooden floor, the weight of his M1A1 and prototype harness familiar and comforting.

"SSG Mercer," CPT Walker said, looking up from the maps. "Glad you could make it. We've got a mission that requires your expertise—and your team's."

Jack nodded, folding his hands behind his back. "Yes, sir. What's the objective?"

CPT Walker gestured to a marked section deep behind the enemy lines. "Recon patrol. We need eyes on the North Korean HQ in this sector. We don't have any intel on the defenses, troop strength, or command structure. That's where your team comes in."

Jack leaned over the map, studying the terrain. Hills, river crossings, and sparse villages dotted the path they'd have to take. "How deep?" he asked.

CPT Walker's gaze darkened. "Deep enough that it's risky. Company patrols have already been sent into this area, and we've taken losses. Men we can't replace easily. That's why this has to be precise, Mercer. Minimal contact, maximum intel. Your judgment will decide who comes back."

Jack exhaled slowly, weighing the challenge. "Understood, sir. We'll move carefully, observe, and report. Minimal engagement, maximum intel."

CPT Walker nodded. "I've seen what you've done with your team, Mercer. I trust you to get it done without raising alarms. First light tomorrow. I want your team briefed, gear ready, and contingency plans in place. This isn't a patrol for rookies, this is deep behind enemy lines, Mercer. Lives will depend on your judgment."

Jack's hand rested briefly on the silver cross in his pocket, a silent reminder of everything he had to return to and the responsibility he carried for those under his command. "We won't let you down, sir."

"See that you don't," CPT Walker said, voice firm. "Dismissed."

Jack left the command center, already running scenarios in his mind. Terrain, enemy positions, patrol formations, extraction routes. He would need every ounce of experience, every lesson from Normandy, every second of training since arriving in Japan. And his team would need him sharp, focused, and ready for anything.

The briefing tent beside the company command post was lit by a single swaying lantern, its glow throwing long shadows across the canvas walls. The air inside was thick with the scent of damp canvas, cigarette smoke, and the faint musk of oiled weapons. Jack's squad sat around a folding field table, with maps and grease pencils spread out before them. Outside, the muffled rumble of trucks and the distant bark of orders reminded them the whole camp was alive with preparations.

SSG Mercer stood at the head of the table, helmet resting beside him. The men waited, silent but alert. Davis had his notebook open, ready to jot down notes. Kim leaned forward, arms folded across the table. Reilly tapped his boot nervously, and Hill flicked ash from the stub of a Lucky into an empty tin.

Jack cleared his throat. "Listen up. I just came from CPT Walker. We've been tasked with our first mission here in Korea." He laid a map flat out on the table and smoothed it with one hand. "Deep recon. Our job is to slip behind the lines, identify where the enemy HQ is operating from, and track any troop movements in the sector. No firefights, no flag waving—we're shadows on this one."

Reilly exhaled through his teeth. "So, we're walking into a hornet's nest, SSG Mercer?"

"That's about the size of it," Jack replied evenly. "Other patrols have tried, and they've taken losses. Good men didn't come back. This isn't a milk run. It's dangerous, and CPT Walker made it clear he doesn't want any more names added to the list."

The lantern hissed softly, filling the silence that followed.

Kim finally broke it. "We can handle it. Long as we're tighter than the other patrols."

"Exactly," Jack said, locking eyes with him. "That's why we've spent weeks training harder than anyone else. Noise discipline, movement drills, fieldcraft. We're going to use all of it." He tapped the map again. "Hill, you're on point—you see first, you warn first. Reilly, you're rear security. Kim, you will stay with me on navigation. Davis—you're my anchor. If anything happens to me, you make the call."

"Roger, SSG Mercer," Davis answered, pencil scratching notes in his book.

Jack let his gaze sweep the table. "This is recon, not glory hunting. If we make contact, we break. If we're compromised, we melt away. Our job is to come back with answers, not bodies."

The men nodded, each one sober, the weight of the mission settling on them. Even Reilly, usually quick with a wisecrack, kept quiet.

Jack leaned in, his voice low but sharp. "Normandy taught me what happens when plans go bad. I won't see that happen again. You follow orders, you trust each other, and we come back together. Understood?"

The reply was steady, in unison: "Roger, SSG Mercer." As the team looked at him with a questionable look on their faces. The Normandy comment caught them all off guard.

Jack straightened, picking up his helmet. "Gear check in thirty minutes. Then we run through our movement one more time before dawn. Tomorrow, boys, we're ghosts."

The squad rose, gathering maps and gear, the scrape of chairs and clatter of webbing loud in the confined tent. As they filed out into the cold night air, Jack lingered a moment, staring down at the lantern lit map. CPT Walker's warning echoed in his mind—*losses*. He swore to himself his team wouldn't be added to that list.

The cold December air cut sharply as the squad gathered outside the briefing tent. A low lantern and a few burning stubs of candles lit the patch of dirt where Jack had hastily built a terrain model—a ridgeline scratched in with his knife, rocks marking villages, and sticks laid out as roads. The men circled around, rifles slung, shadows long and flickering against the canvas walls behind them.

SSG Mercer crouched low, pointing to the crude model. "This ridge here—we'll cross it just before dawn, move into this valley. Hill, that's your lane. Every tree, every shadow, you call it. No surprises."

"Yes, SSG Mercer," Hill replied, eyes locked on the model like it was gospel.

Jack shifted his finger to the little stack of pebbles representing a cluster of farmhouses. "Kim, this is where we'll break for the first hide. You and I will map the approach, then we hole up here till dark. No fires, no noise. Just eyes and ears."

Kim nodded. "Got it, SSG Mercer. Nothing gets past us."

"Reilly," Jack continued, tapping the rear of the model. "You own the tail. If anything comes up behind us, we'll never hear it first you will. You're the tripwire. Don't miss it."

Reilly cracked a grin, but it didn't reach his eyes. "Won't miss a thing, SSG Mercer."

Finally, Jack pointed to the far edge of the dirt, where a cluster of sticks marked enemy positions from intel. "Davis—you keep the book.

Patrol log, compass checks, times, routes. If we get cut off, you'll be the one who can walk us home."

"Yes, SSG Mercer," Davis answered firmly, scribbling in his notebook even now.

Jack stood, dusting dirt from his gloves. "Now, we walk it."

For the next hour, he drilled them hard. Each man took turns stepping through the model, explaining their role as Jack quizzed them—sometimes stopping mid-sentence to throw a curve.

"You just spotted movement on the flank. Hill, what do you do?"

"Signal back—silent hand, no voice. Freeze the line until cleared."

"Good. Reilly—you hear footsteps behind us, too close. What's your move?"

"Drop low, eyes back. Silent alert forward. If they get too close, I peel off and draw 'em away."

"Correct."

Repeatedly, they rehearsed until the responses were instinct, not thought. The men moved with increasing confidence, their earlier nerves sharpening into quiet readiness.

When at last Jack called halt, the squad stood in the bitter night air, their breath fogging in the lantern light. The terrain model sat between them, no longer a crude scratch of dirt and sticks, but the battlefield burned into their minds.

Jack looked them over, his voice steady, low. "Tomorrow, we step off. You've got the plan. You've got each other. Stick to both, and we come back."

"Roger, SSG Mercer," they echoed together, the words carrying weight in the frozen night.

The tent was dim and cold, canvas snapping lightly in the wind outside. One by one, the men settled onto their bedrolls. Davis stretched out with a low groan, muttering something about finally being off his feet. Hill tugged his blanket up to his chin, staring at the tent's ceiling as though he might will himself to sleep. Kim lay on his back, hands folded over his chest, eyes closed but clearly still awake. Reilly shifted endlessly, rolling from side to side until Davis told him to "knock it off" in a tired growl.

Jack lay still, listening to their voices fade into silence. To anyone looking, he seemed calm, but his mind was anything but. He thought first of Patricia—her smile, her steady gaze, the way she had made him feel like he wasn't alone in the world anymore. That warmth carried him only so far before the darker memories took hold.

Normandy 1944.

The harness cut into his shoulders again, wind whipping past as tracers clawed the night sky. He could see them even now Rizzo and his fire team drifting nearby, their parachutes blossoming against the darkness. For one fleeting moment, Jack had believed they might hit the ground together, fighting side by side. Then came the tearing roar of a German MG-34.

The weapon spat death from the hedgerows, its staccato fire ripping through silk and flesh. Rizzo's canopy shredded mid-air, his body jerking as rounds found him and his men. They hung limp before they ever touched soil, their parachutes collapsing into bloody tatters that swayed above Normandy's fields. Jack had pulled hard on his risers, drifting away with a hollow scream stuck in his throat—helpless to do anything but watch.

Back in the Korean night, Jack blinked against the memory. He turned his head slightly, listening to the breaths of the men around him—his men now. Davis, Hill, Kim, and Reilly. Young. Green. Trusting him to lead them into hell and back.

Jack clenched his fists. Not again. He would not let history repeat itself. He would bring them home—whatever it took.

The morning came gray and brittle, a low mist clinging to the camp like a shroud. The Rangers moved with quiet efficiency, strapping down gear and checking weapons. Canvas packs thumped against the frozen ground, canteens sloshed, metal clips snapped into place.

SSG Mercer moved among them, inspecting every buckle, every magazine, every strap. Davis was cinching down his rucksack, muttering about the weight; Hill worked the bolt of his rifle until it clicked smooth; Kim methodically checked the alignment of his compass and map; Reilly fussed with his boots, cursing under his breath when the laces froze stiff.

Jack raised his voice just enough for the squad to hear. "Alright packs on. Jump test. Now."

The squad slung their rucksacks onto their backs; rifles secured and began hopping in place like recruits on a drill field. At once the air filled with clinks, rattles, and squeaks. Canteens slapped against canvas, straps creaked, metal kissed metal.

Jack joined them, his own pack heavy and solid against his shoulders. He jumped, landing hard, listening to the faint rattle from one of his ammo pouches. He nodded. "Good. That's why we do it here and not out there. Strip it down. Tape it up. Make it silent. If it makes a sound, you fix it—or you don't carry it."

The team set to work, grumbling but focused. Canteens were wrapped in rags, buckles padded, loose straps bound tight with tape. When they were finished, Jack had them repeat the test. This time, the squad jumped in unison. The only sound was boots striking frozen earth.

Jack allowed himself the smallest smile. *Better. Quieter. Alive.*

The memory of Rizzo and his fire team still burned hot in Jack's chest, but instead of dragging him down, it hardened into steel. He would not lose these men. Not to an ambush. Not to careless noise. Not to his own failure.

As he tightened the strap on his own harness, his hand brushed the familiar worn steel of his paratrooper's M1A1. Old gear, new war—but it was his. He adjusted the prototype rig slung across his chest, earning a sidelong grin from Davis.

"Never thought I'd be following a man half GI, half museum exhibit," Davis joked.

Jack gave him a thin smile, though his eyes stayed cold. "Doesn't matter what it looks like. All that matters is it works. And we make it back."

The men grew quiet at that, the weight of his words settling over them heavier than their gear.

Inside, Jack repeated the vow he'd made the night before. *Not again. Not them.*

He slung his pack into place and straightened. "Alright, Misfits," he said. "Form up. Time to earn our keep."

The ward was dim except for the harsh glow of overhead lights and the steady rhythm of footsteps and whispered orders. Patricia moved briskly between the rows of cots, her hands steady though her heart was not. Bandaged men shifted and groaned, some caught in fever, others staring into the distance with eyes that had seen too much.

She adjusted an IV line for a young private whose leg was swathed in bloody dressings, then smoothed the blanket over his trembling hands. "You're safe now," she whispered, though she wasn't sure if he heard her.

The hospital reeked of antiseptic and iron, of sweat and pain. Patricia had grown used to the sounds—murmured prayers, sudden cries, the soft rasp of nurses and orderlies moving from one bed to the next. What she hadn't grown used to was the hollow ache in her chest every time her thoughts drifted to Jack.

"Where is he now?" she wondered. *"Already across the sea?" "Already in the cold hills of Korea?"*

She tried to focus on her work, but the image of him haunted her—the steady way he'd held her hand, the fire in his eyes when he spoke of protecting his men, the quiet strength in his voice when he asked her to marry him. She could still feel the warmth of that moment, though it was wrapped in secrecy.

Pausing at the foot of a cot, she rested her hands on the rail and let her gaze slip unfocused. *I wish I could be near him... even just once more. To tell him again that he's not alone. That we're in this together, no matter how far apart.*

She blinked hard, drawing in breath before moving on to the next soldier, her face composed but her heart aching. The war was pulling them in opposite directions—her bound to the wounded who returned, him carried deeper into the fight.

And yet, somewhere in her soul, she clung to the belief that fate wouldn't be so cruel as to tear them apart completely.

For a moment, she closed her eyes and imagined Jack standing beside her, guiding his men through another perilous mission, keeping them alive as he always had. And in that instant, the weight of her secret became lighter, held silently between them across the miles.

Opening her eyes, she continued her rounds, each movement precise, each touch careful. She could not be with him—not now—but she could hold onto the hope that somehow, they would meet again.

Patricia moved along the rows of cots, adjusting bandages and checking IVs, her hands steady but her mind still on Jack. She imagined him in the cold hills of Korea, leading his men, keeping them alive the way he always did. A pang of longing struck her chest—she wished she could be there, even just to see him one last time before the unknown pulled him into danger.

She paused at the foot of a cot, letting her eyes close for a moment, and in her mind she whispered a promise: *Come back to me, Jack. Come back.*

The image faded, and the ward returned to quiet murmurs and the faint scrape of shoes on linoleum. Far away, Jack was already moving.

He checked the final adjustments on his gear, making sure every strap was tight and every pouch silent. The team followed his lead, jumping lightly in place to test for rattles, each man mimicking his motions until the packs and equipment were as silent as the snow-dusted ground around them.

Jack's eyes swept over his men—Davis, Hill, Kim, Reilly. Each one focused, alert, carrying the weight of the mission and the unspoken trust that they would all get through this together.

The faint memory of Normandy flickered at the edges of his mind—the MG-34, Rizzo's fire team falling before they hit the ground—but Jack forced it down, letting it sharpen rather than paralyze him. *Not again,* he told himself. *Not them.*

He slung his pack into place and checked his M1A1, feeling the familiar steel against his chest.

"Alright, Misfits," he said, voice low but firm. "Form up. Time to move."

The men fell into line, silent shadows in the gray morning light, each step measured, each movement precise. Jack's resolve was as cold and hard as the steel he carried.

The wire loomed ahead, its cold metal glinting faintly in the gray dawn. Jack's hand hovered over the gate release, checking the alignment of his pack and the tension of his straps one last time. Around him, Davis, Hill, Kim, and Reilly mirrored his motions, eyes sharp, muscles coiled like springs.

Jack gave a quiet nod. "Move."

They slipped forward, one at a time, crawling through the narrow opening, boots barely making a sound on the frozen earth. The wire clinked faintly against their gear, but they adjusted, testing the spacing and timing to keep it as silent as possible.

Jack led the way, scanning the ridge lines, the valleys, every shadow and outline. He could feel the weight of responsibility in every step—the lives of his men, the memory of those he'd lost, the oath to return them all home.

The last man cleared the wire, and for a heartbeat, the world seemed suspended. The camp behind them was quiet, still, unaware of the small team that had just vanished into the gray morning.

Jack exhaled, letting the tension drain slightly, and whispered to himself, *not again. Not them.*

Beyond the wire, the hills of Korea awaited. Every step would be measured. Every decision mattered. And the patrol had only just begun.

Chapter 13: Shadows and Grids

The cold bit deeper once they were beyond the wire, as if the land itself wanted to remind them they weren't welcome. Frosted grass crunched softly beneath their boots, muffled by the careful pacing Jack demanded. The hills ahead rose like jagged teeth, gray shapes lost in mist, and every step forward seemed to swallow them further from safety.

Hill took point, his silhouette barely more than a darker shadow moving against the terrain. Jack followed close, eyes flicking between the ridgeline and the valley floor, mind mapping every fold of the land. Behind him, Kim checked their direction against his compass, Davis scribbled times and bearings into his notebook, and Reilly moved with restless energy at the rear, head turning like a wolf scenting danger.

The silence was oppressive. No trucks, no engines, no voices. Just the rasp of breath and the soft squeak of leather under strain. Even the wind seemed to hold its breath.

Jack's senses sharpened in the quiet. Every twig, every distant echo sets his nerves alight. Normandy still lingered like a ghost at his shoulder, whispering warnings of what came when vigilance faltered. He forced it down, channeling the unease into focus. His men were counting on him.

An owl hooted somewhere along the tree line, the sound startling in the dead stillness. Hill froze and raised his hand, the line halting instantly. Jack slid up beside him, scanning the slope ahead. Dark shapes clustered near a half-collapsed barn at the valley's edge. For a heartbeat, Jack's gut clenched—enemy patrol?

Then one of the shapes shifted, bleating softly. Goats. Just goats.

Jack breathed out slowly, motioning Hill forward again. The line crept on, silent shadows vanishing into the Korean hills.

The hills thickened as dawn bled into day, the mist burning away to reveal terraces of frozen fields and lonely farmhouses abandoned to the war. The Misfits moved like ghosts, single file, each man's eyes sweeping for threats. Every step deeper into the countryside stretched the tension tighter.

Hill suddenly dropped into a crouch, his hand shooting up. The line froze. Jack slid forward, following Hill's finger toward a dirt road cutting across the valley floor.

A column of North Korean infantry was moving along it—thirty men at least, rifles slung, their breath clouding in the cold. A couple wagons creaked behind them, piled high with crates. An officer barked something in clipped syllables, and the men quickened their pace.

The squad sank lower into the brush, hearts thudding in their ears. Reilly pressed his face into the dirt, whispering under his breath, "Christ, that's a lot of them…"

"Quiet," Jack hissed, eyes locked on the column. One wrong move—one glint of steel, one careless sound—and the patrol would have them pinned in open ground.

Minutes dragged like hours as the enemy passed. Finally, the last man disappeared behind a rise. Jack counted to thirty before raising his fist and signaling forward. The Misfits rose silently, resuming their path, each man carrying the weight of what they had just avoided.

By midday they were deep into the hills. The terrain grew harsher—narrow ridges, gullies tangled with brush, frozen streams that cracked underfoot. Sweat mixed with the cold, chilling them despite their layers.

It was Hill again who spotted movement ahead—two North Korean soldiers cresting a rise, rifles slung casually, talking in low voices. The Misfits dropped instantly into the rocks, rifles ready. Jack's hand shot up—*hold*.

The two soldiers stopped not twenty yards away, one lighting a cigarette while the other relieved himself against a boulder. The acrid smoke drifted downwind, close enough Jack could taste it. His thumb brushed the safety of his M1A1, every nerve urging him to act, but he forced himself still.

Minutes stretched. The soldiers laughed at some crude joke, then finally moved on, footsteps fading into the valley.

Only when the air cleared did Jack lower his hand. He turned to his men, his voice a bare whisper. "That close is too close. Noise, light, movement none of it. We're ghosts, remember?"

The men nodded, faces pale but steady.

As the sun dipped lower, the team pushed deeper into hostile ground. Patrols passed within earshot, boots crunching on frozen earth, foreign words carried on the wind. The Misfits stayed low, moving with the patience of hunters, hearts pounding with every close call.

By the time the sky turned orange over the hills, Jack's nerves felt strung to the breaking point—but they were still alive, still unseen. And the enemy HQ was somewhere ahead, waiting to be found.

The ridgeline dropped into a wide valley; its floor cut with frozen paddies and dirt tracks that converged on a cluster of buildings. From their perch, half-hidden by brush and frost, the Misfits could see it all: trucks parked under netting, radio antennae jutting from a farmhouse roof, and guards pacing in regular intervals.

Jack raised his field glasses, adjusting the focus until the scene sharpened. North Korean sentries in quilted jackets trudged along the perimeter, stamping their feet against the cold. At least two machine gun nests were dug into berms overlooking the road approach. A pair of canvas-topped trucks sat backed against the barn, crates stacked high beside them—ammunition or rations, he couldn't tell.

"Christ," Reilly muttered, low and tight. "That's a whole damn hornet's nest."

"Quiet," Jack snapped softly, never lowering the glasses. He tracked a pair of officers emerging from the farmhouse, and maps clutched under their arms. They spoke briefly to a radio operator before disappearing inside again.

Jack's mind worked fast—HQ building, signals station, supply points, defensive placements. He shifted the glasses, sweeping wider. To the north, a camouflaged path led into the hills—likely an approach route for reserves or supply. To the east, a crude pen held a string of pack mules. To the west, smoke rose from hidden cookfires.

He handed the glasses to Kim, who adjusted his compass and bearings, whispering, "Marking grid square… defensive emplacements… possible reserves along this track." His pencil scratched quickly across the notebook Davis held steady.

They inched their way around the ridge, stopping often to observe from new angles. From the southern slope they saw more—an artillery piece dug in under netting, its barrel angled skyward, guarded by a half dozen men. Davis recorded each position, drawing a rough sketch map with times, compass checks, and unit estimates.

Hill, belly down in the frost, pointed toward the tree line near the compound. "Patrol pattern. Two-man teams every ten minutes. They loop the perimeter clockwise."

Jack nodded. "Good eye. That's our way in and out if higher decides to hit this place."

For hours they watched in silence, piecing together the puzzle. Numbers, weapons, supply lines. Each detail mattered. Each one could save lives when CPT Walker and higher HQ planned their strike.

As dusk began to creep in, Jack finally signaled the team to withdraw. They slid back into the cover of the ridge, moving slowly and silent until the enemy compound was out of sight. Only then did Jack allow them to breathe more easily.

Reilly let out a low whistle. "If that doesn't keep CPT Walker's map boys busy, nothing will."

Davis snapped his notebook shut. "We've got strength estimates, supply markers, patrol routes, comms. It's enough to build a whole picture."

Jack adjusted his harness, eyes still scanning the horizon. "Then we did our job. Now we get it home."

The Misfits moved off into the gathering dark, shadows blending with the hills. Behind them, the enemy HQ churned with activity, unaware it had just been mapped down to the last shovel of dirt.

The Misfits moved slowly, deliberate, shadows stretching thin under the rising moon. Jack had steered them onto a different route—never backtracking, never giving the enemy a chance to pick up their trail. The ridge carried them south, winding through scrub and frozen gullies, the silence broken only by their controlled breathing and the crunch of frost beneath careful boots.

Halfway through the night, Hill froze at the front of the line, dropping low and raising a fist. Jack crept forward, heart hammering, until he saw it too: faint orange glows through the trees ahead, the dull clank of metal on metal, and the low murmur of voices.

The squad eased into the brush bellies to the ground. Through a screen of pines, the scene opened before them. A hidden artillery park spread across a clear stretch of ground. Two 122mm Soviet-made howitzers crouched like beasts, their barrels angled toward the southern valleys. Camouflage netting sagged above them, though the moonlight still picked out the glint of steel. Around the guns, crews moved with weary efficiency—stacking shells in neat rows, checking elevation, stoking cookfires that hissed and smoked.

Jack's gut tightened. This wasn't some temporary setup. The positions were dug in—ammo pits, command dugout, even telephone lines running back into the hills. This battery was meant to stay, and it was pointed squarely at the Allied lines.

He signaled the men to fan out into observation positions. Davis unfolded his notebook once more, Kim whispering compass bearings while Hill and Reilly kept eyes on the perimeter. Jack pulled out his own field glasses, scanning for details.

"Two guns," Kim breathed. "Crew strength—at least a dozen per piece. Plus, support. Call it thirty, maybe more."

"Ammo supply's heavy," Davis murmured, sketching the stacks. "I count… fifty rounds staged, easy. More in the dugouts."

Hill pointed with his chin toward the tree line opposite. "Trucks. Three of 'em. Covered. Supply line still active."

Jack noted each point—guns, ammo, transport, comms. He shifted the glasses toward the far side of the position, catching the outline of a lean-to with a field radio antenna. "Fire direction center," he whispered. "That's their brain."

Reilly exhaled low. "We could knock 'em out if we had the kit. Whole sector would breathe easier."

Jack shook his head. "That's not our fight tonight. Our job is to bring CPT Walker the picture. Higher will decide how to break it."

For nearly an hour they watched, recording every routine the change of shifts on the guns, the route of sentries, the time it took for runners to move between the battery and the command post. Davis's pencil scratched furiously, the page filled with arcs, arrows, and numbers.

Finally, Jack eased back into the shadows, giving the signal to withdraw. The Misfits regrouped in silence, their faces tight with the weight of what they'd just uncovered.

Jack glanced once more toward the enemy artillery, the thunder it promised still silent for now. "Alright," he whispered. "We've got enough to choke CPT Walker's map table. Now we get it home."

The squad slipped back into the night, each step heavier with the burden of intel and the knowledge that discovery now would mean disaster—not just for them, but for every man depending on the lines they were racing to protect.

The march back became a crawl. Every hundred meters seemed to reveal something new— another glimpse of just how thick the enemy had dug in.

First it was a cluster of dugouts, half-buried under camo netting, soldiers' breath puffing in the cold night air as they huddled around a lantern. Then, not a hundred yards further, Jack spotted the dark outline of a mortar pit, four tubes pointed skyward, crates of rounds stacked close at hand.

The Misfits hugged the ground, inching past, every heartbeat louder than the muffled voices of the enemy below.

They slid into a drainage gully, only to freeze again when Hill raised a clenched fist. Jack crept forward, eyes narrowing. Across the slope, a pair of machine gun nests were dug in with sandbags, each one covering the valley approach. Sentinels shifted nervously at their posts, rifles slung loose, but their eyes never stopped scanning the darkness.

Jack cursed silently. The deeper they pushed, the more he realized—they weren't just behind enemy lines. They were in the middle of a fortified zone.

Kim whispered the thought that hung in all their minds: "It's like we're walking through their whole damn army."

"Quiet," Jack breathed. "Keep moving. Log it all."

Davis scribbled furiously, his notebook now fat with bearings, numbers, and sketches. Every page was a map of death—gun pits, patrol routes, supply caches. If they made it back, CPT Walker would have the clearest picture yet of the enemy's strength. If.

They crawled another stretch, the ridge giving way to a hollow. Jack's gut sank when he saw it: another artillery park, this one smaller, but with fresh tracks leading out of it. Trucks loaded with shells. Men hauling crates under lanternlight. A full logistics line running at night.

Reilly muttered under his breath, "Jesus, they've got enough here to blow the whole front wide open."

Jack gave him a sharp look, then scanned the sector one last time. Every discovery raised the stakes higher. The weight of responsibility pressed on his shoulders like stone. One slip—one sound too loud, one step too fast—and they'd vanish in a storm of rifle fire, their hard-won intel buried with them.

Pulling them back into cover, Jack whispered, "Alright, Misfits. We've seen enough. No more detours. We move slowly, we move smartly, and we get this home. Every step we take carries a hundred lives with it now. Don't forget that."

The squad nodded, silent and pale, then followed him back into the dark. Every shadow seemed to move, every tree seemed to whisper, but they pressed on—carrying with them the knowledge that the hills of Korea were far more dangerous than anyone in CPT Walker's tent could have guessed.

The hills thinned into scrub as the Misfits crept closer to the wire. Jack knew the distance by instinct—less than half a kilometer now. Close enough to hear the faint crackle of Allied radios in the night air, close enough that a single wrong move could see them cut down by friendly fire.

He raised a fist, signaling haLT The men dropped into the frost, packs sliding quietly from their shoulders. Jack eased his own gear down, pulling the radio handset free. Davis was already flipping through his notebook, ready to pass authentication codes.

Jack pressed the handset to his lips, whispering, "Lancer Six to Outpost, over."

Static hissed in reply. He adjusted the dial, tried again. "Lancer Six to Outpost, authenticate Whiskey-Four. Requesting recognition, over."

Before the crackle of an answer came, Hill tensed beside him. His hand shot up in a warning fist, finger pointing downslope.

Movement.

Two figures slipped through the brush less than fifty yards away. Both wore quilted jackets that melted into the scrub, moving with deliberate care. One carried a set of binoculars, the other kept glancing at a notebook, jotting down bearings and sketches.

Jack's chest tightened. Not a patrol. Recon. These two were charting the wire, cataloguing fields of fire, looking for weaknesses. If they got back with what they'd seen, CPT Walker's line would pay the price.

He slid back to his men, crouching low. His voice was a razor-thin whisper.

"Two scouts. Reconning the line. We can't let them take that intel home. We're taking them alive."

Reilly frowned. "Capture them? Out here?"

Jack's eyes were hard as flint. "That's right. Quiet. No gunfire unless you have no choice. We need them breathing proof of what's moving out here."

The Misfits shifted uneasily but gave quick nods.

Jack pointed, assigning roles. "Hill, swing left. Reilly, wide right. Kim, with me. Davis, hold position—you're our anchor."

The squad melted into the shadows, tightening the noose. The two enemy scouts knelt in the scrub, binoculars trained on the Allied perimeter, oblivious to the danger creeping in around them.

Jack adjusted his grip on the folding stock of his M1A1A1 Paratrooper Carbine, muscles coiled. His heartbeat slowed, focus narrowing to the two men ahead. This wasn't just about survival anymore; this was about turning the tables.

He leaned close to Kim, voice low and calm. "On my mark… we take them."

Jack and Kim slid forward like shadows, their movements deliberate, each step chosen with care. The two enemy scouts were crouched in the brush, one sketching in his notebook, the other glassing the Allied wire. Neither so much as twitched at the sound of the Misfits closing in.

Jack crept within arm's reach, the folding stock of his M1A1A1 Paratrooper Carbine tucked snug against his shoulder. He eased the barrel forward until the cold steel pressed firmly against the back of the scout's head. At the exact same moment, Kim mirrored him—his weapon kissing the skull of the second man.

The scouts froze, bodies rigid. The notebook tumbled into the grass.

"Don't move," Jack hissed, his voice a predator's growl.

From the flanks, Hill and Reilly surged in, swift and silent. Hill ripped the binoculars away while Reilly wrenched the rifle from the scout's hands. Both Misfits moved with practiced efficiency, forcing the men face-down into the dirt. Rope was produced in seconds; wrists yanked behind backs and bound tight. Gags were shoved into mouths, muffling any cry for help before it could be uttered.

About fifty yards back, Davis crouched low beside a rock outcropping, radio handset pressed to his lips. His voice was low and urgent. "Outpost, this is Lancer Six Actual. We have two enemy prisoners, I say again, two prisoners in custody. Request immediate guidance and recognition signal. We're closing the last five-hundred meters. Over."

Static hissed, then the reply came from the operator. Davis scribbled notes quickly, giving a thumbs-up toward the team.

Jack gave a sharp nod. "Let's move."

Hill and Reilly hauled the bound scouts to their feet, gagged and stumbling. Jack and Kim slipped back into formation, every man alert. The wire was close now, but the danger wasn't over. They still had to carry the prisoners the final stretch unseen.

Jack glanced at Kim, then forward toward the faint shimmer of Allied lines. His jaw clenched. *One last push. We get them home, CPT Walker gets proof, and the brass finally knows what we're up against.*

The Misfits moved like ghosts, shadows stretched long and low under the moonlight. Each step was measured, deliberate—Jack's M1A1A1 Paratrooper Carbine tight in his grip, Kim mirroring him at the flank. Hill and Reilly kept the prisoners pressed low between them, rope taut, hands bound behind their backs.

Davis emerged from the brush about fifty yards back, radio in hand, and quietly caught up with the team. "Lancer Six Actual to Outpost—two prisoners secured. ETA five minutes," he whispered, crouching low beside Jack. His presence made the team whole again, all eyes and ears focused.

Jack scanned the terrain, every ridge, clump of brush, and glint of metal a potential threat. The cold night air bit at their faces, but it wasn't the frost that made his chest pound—it was the knowledge that one wrong step, one misread shadow, and the entire operation could erupt into gunfire.

A rustle to the left drew Hill's attention. He froze, eyes narrowed. Jack pressed his hand to his shoulder, signaling caution. Kim swept the area with field glasses.

Nothing. Just the wind whispering through the scrub.

Jack gave the signal to move, and the team shifted forward in tight formation. Every step was slow, silent, purposeful. The prisoners shuffled along, aware of the weapons pressed against their backs but powerless to resist.

Ahead, faint glimmers of wire fencing and dim sentry lights marked the Allied outpost. Relief brushed Jack's mind, but he pushed it down—too many variables still. A flashlight beam, a misaligned guard, a single snap of a twig could trigger disaster.

Reilly murmured, "Eyes open… every step counts."

Jack tightened his grip on the paratrooper carbine. His pulse steadied but every sense screamed alert. The team pressed forward, every movement calculated, each man perfectly attuned to the others.

Fifty meters. Thirty. Twenty.

A sentry stepped into view, flashlight swinging too close. Jack froze, muscles taut as wire. With a subtle hand gesture, he signaled Davis, who keyed the radio. A faint crackle confirmed recognition from the outpost. The beam swept past harmlessly.

Jack exhaled, the tension coiling in his chest like steel. One last stretch remained. The team moved as a single unit, prisoners pressed low, rifles and carbine ready. Every shadow could conceal a threat; every step mattered.

Finally, the Misfits crossed the last stretch of no-man's land. Wire underfoot, the silhouettes of friendly guards ahead. Davis dropped the radio and slid into formation. Hill and Reilly guided the prisoners forward, keeping them low and quiet.

Jack scanned the perimeter one last time, muscles still coiled, eyes sharp. The enemy scouts were now in friendly hands. The team had executed flawlessly silent, precise, alive.

Jack allowed himself a grim, tight smile. *Not again. Not them.*

The Misfits trudged the final few yards into the outpost, the prisoners guided carefully between Hill and Reilly. The crisp night air was thick with tension, every soldier alert as they approached the G2 section.

Two intelligence officers stepped forward, hands raised slightly in acknowledgment of the secured captives. Jack handed the ropes to Hill and Reilly, who quickly transferred the prisoners into the custody of the G2 personnel. The scouts were led away, their movements restrained, eyes darting, muffled sounds struggling against the gags.

Jack turned, leading his team toward the command tent. CPT Walker was already waiting, maps spread across the table, lantern light glinting off the polished surface. Jack unrolled his field sketches, pointing to the grid coordinates, trenches, and positions.

"We mapped the HQ in detail," Jack began, his voice calm but edged with urgency. "Enemy strength, firing positions, patrol routes. Mortar pits, supply caches, and an artillery unit thirty-five meters northwest of the primary ridge. Every encampment we came across is logged in this grid. I've cross-referenced distances and fields of fire. We know patrol routines, troop strength estimates, and approach routes for each site."

CPT Walker leaned over, eyes sharp, following Jack's finger across the grid. "Mortars?" he asked.

"Three emplacements," Kim replied, nodding. "Each camouflaged. We noted reload times, firing patterns, and nearest observation points. They're all within range of the artillery coordinates we gathered."

Hill pointed to a set of coordinates along the approach route. "Every cluster of enemy troops we saw is marked. Movements, schedules, and estimated numbers. Nothing is guesswork."

Reilly added, "We even tracked the minor patrols—two to four men—moving between the main encampments. If artillery is called in, these positions can be neutralized or disrupted first."

Davis passed CPT Walker the notebook filled with sketches, bearings, and timing logs. "Every observation timestamped. Compass points, distances, terrain notes. Should be enough for HQ to plot fire missions with confidence."

CPT Walker leaned back, his eyes scanning the table. The silence in the tent was heavy with respect. Finally, he let out a low whistle.

"Outstanding work, Misfits," he said, voice firm but carrying pride. "This is exactly what I was hoping for. You've not only survived behind enemy lines you've brought back actionable intelligence that can save lives and take out key positions. I'll be calling higher HQ to coordinate artillery strikes on every one of these targets you've mapped. They're going to hit hard, and it's all because of the precision and discipline you've shown out there."

Jack nodded, his jaw tight, relief mingling with exhaustion. CPT Walker's gaze swept the team. "Now, go eat. Get some rest. You've earned it. But keep your head clear we've got another mission lined up. I'll brief you around 2000 hours after you've recovered and fueled up. And Misfits… I expect the same performance."

The team allowed themselves small smiles, the tension of the past hours easing slightly. For now, they had delivered their hard-won intelligence safely into the hands of command, set the stage for a decisive strike, and survived the night together.

Jack tucked his sketches and notebook away, slinging his M1A1A1 Paratrooper Carbine across his shoulder. "Let's get something warm in our stomachs," he murmured to the squad. Kim, Hill, Reilly, and Davis followed, silent but proud, knowing they had just earned the respect of CPT Walker—and written another line in their own history behind enemy lines.

Chapter 14: Coordinates of War

The small mess tent was quiet except for the occasional scrape of metal against canvas as the Misfits unpacked their gear and settled into the warm glow of a lantern. The air smelled faintly of coffee and reheated rations, a welcome reprieve from the frozen hills outside.

Jack leaned back against a crate, the M1A1A1 Paratrooper Carbine resting across his lap, and surveyed his team. Kim cleaned his rifle methodically, Hill rubbed at a sore shoulder, Reilly absentmindedly adjusted the rope bindings he'd helped secure the POW's with earlier, and Davis scribbled furiously in his notebook, organizing notes from the mission.

"Alright," Jack began, voice low but firm. "Let's go over what we did well—and what we need to tighten up for next time. Nobody's perfect, and we're only going to get better if we're honest with each other."

Hill leaned forward. "Movement. I think we kept tight formations and silent as ghosts. Even with the terrain and the frost, I didn't hear anyone slip up."

"Agreed," Kim added. "We stayed on the grid, we checked each other, and we weren't rushed. That's why we didn't get spotted until the last stretch. But..." His eyes flicked to Jack. "...we need to make sure our prisoners don't struggle. That gag and tie method works, but a second set of hands or a quicker technique would save us time."

Reilly nodded. "And timing. The last hundred meters had me on edge. We need to rehearse the approach closer to the wire so we're smoother. Heart rates don't lie—it's easier to slip if we're not counting every second."

Davis flipped a page in his notebook. "Intel gathering was solid. The maps, the coordinates, everything is logged. But I think I could've been faster with the compass bearings and cross-checking with terrain features. Next time I need to be a step ahead, not catching up."

Jack absorbed it all, nodding slowly. "Good points. But remember, nothing broke. We got the prisoners, mapped the HQ, the artillery, the mortars, all the enemy encampments—and we did it quietly. That's a win. Still, we tighten the final approach, improve prisoner control, and speed up our log entries. That's our margin for next time."

Kim leaned back, a small smile tugging at his lips. "And we still all made it back in one piece. That's got to count for something."

Jack allowed a faint grin. "It does. But keep in mind—this is the baseline now. Every mission, we improve. Every mission, we survive. And every mission, we make CPT Walker proud."

The men exchanged quiet nods, the camaraderie and shared tension of the night settling in like a shield. Outside, the wind whistled through the hills, but inside the tent, they were focused, sharp, and ready to absorb the next set of orders.

Jack glanced at his team. "Get a bite, hydrate, stretch. CPT Walker's going to brief us at 2000 hours. We eat, we rest, then we step into whatever comes next."

The cramped warmth of the sleeping tent pressed around Jack as he lay on his bedroll, the faint scent of canvas mingling with sweat and the lingering cold of the hills outside. His team was scattered nearby—Kim adjusting the strap on his rifle, Hill quietly rubbing at a sore shoulder, Reilly shifting to find a comfortable position, and Davis flipping through his notes from the mission.

Jack closed his eyes, letting the sounds of the present fade. The cold wind and distant gunfire of Normandy replaced the rustle of canvas. He was crouched behind a hedgerow with Sam, scanning the fields for German positions. The rifles between them were ready but silent, every shadow a potential threat.

"Eyes on the ridge," Sam whispered, pointing toward the orchard. "Looks quiet… but quiet doesn't mean safe."

Jack nodded, adjusting the scope on his weapon. "We map, we watch, we report. Don't engage unless necessary."

A patrol of German soldiers crept cautiously through the fields beyond. Jack traced their paths in the dirt with a stick, noting positions, approach routes, and possible ambush points.

"Reinforcements could hit here within twenty minutes if they detect anyone," Sam murmured. "We need to mark the crossings for HQ—get the artillery coordinates down before anyone notices us."

Jack crouched lower, notebook in hand, the memory of the MG-34 fire earlier that day stabbing at him—the way Rizzo and his men had fallen, helpless in the open.

"Focus, Jack," Sam said softly. "We survive by being methodical. That's how we stay alive."

Jack's jaw tightened. "I know. I won't forget it. Not then, not now."

The memory of mapping enemy positions, noting every trench, machine gun nest, and patrol route, played across his mind like a shadow of the past. Rain slicked the hedgerows, mud clinging to their boots, but the discipline, the silent coordination, and the trust in Sam had kept them alive.

Back in the tent, Jack's hand brushed the worn steel of his M1A1A1 Paratrooper Carbine resting beside him. Normandy had taught him lessons that Korea now demanded—every careful step, every measured decision, every second of vigilance.

He opened his eyes and glanced at his team, scattered across the tent in various states of rest. *This time, I won't let history repeat itself,* he thought. *Not them.*

Jack blinked, shaking off the shadows of Normandy. The rustle of the sleeping tent came back into focus—the muted snores, the occasional shift of boots against canvas. Kim leaned against his pack, eyes half-closed but alert; Hill rubbed his shoulder, muttering about "old bones in a young war"; Reilly fiddled with his boots; Davis shut his notebook with a snap, his mind never far from the next radio call.

The flap of the tent opened, letting in a gust of icy air. CPT Walker stepped inside, his presence immediately sharpening the atmosphere. He carried a map tube under one arm and a steaming tin mug in the other.

"On your feet, Misfits," CPT Walker said, his voice low but carrying. "I know you just got back, but higher HQ isn't giving us much breathing room."

The team stirred, dragging themselves upright. Jack rose last, slipping the strap of his carbine across his chest and standing at ease.

Walker unrolled a map onto the nearest crate, weighting it with his mug. His finger tapped a spot east of their current position. "January 2nd, 1951. That's when the next op goes down. Command wants fresh eyes on these hills here. Intel suggests the enemy is massing more than just infantry—possible armor staging areas, maybe even long-range artillery. We can't confirm it yet."

Jack and his men leaned in closer, the lines and grids on the map eerily familiar to Jack after the recon they'd just pulled off.

Walker's gaze swept the team. "You'll move before dawn, January 2nd. Same rules: avoid contact unless you have no choice. Recon, mark positions, and bring me back a full picture. The last mission you handed me was pure gold. Artillery's already been hammering those targets. Higher HQ is impressed. Let's keep it that way."

Hill gave a short nod. "Sir, what's the exfil?"

Walker tapped another point on the map. "Friendly lines will be here. Same precautions as last time—radio in before you approach, or the boys will cut you down without blinking."

Reilly smirked faintly, trying to mask the tension. "Always nice to know our own guys are just as dangerous as the enemy."

Walker didn't return the smile. He looked to Jack. "Lancer Six, you've got the lead. Train your men, get some rest, and be ready. We step into the hills again in less than thirty-six hours. Understood?"

Jack met his eyes and gave a single nod. "Understood, CPT Walker."

"Good. Then get what sleep you can. You'll need it."

With that, Walker rolled up his map and left as quickly as he'd come, the cold air following him out. The tent fell silent for a moment, only the distant sound of artillery echoing through the night.

Jack sat back down on his bedroll, his thoughts heavy but steady. Normandy was behind him. Korea was here. And January 2nd, 1951, was already looming ahead.

The lantern in the center of the tent cast a dim, yellow glow over the unfolded map. The Misfits were crouched close, shoulders nearly touching as the cold crept in from outside. Jack used the butt of his pencil to trace the terrain—hills, valleys, and ridgelines marked in neat blue grid squares.

"Alright," Jack said, his voice calm but deliberate. "We step off before first light, January 2nd. The route takes us along this ridgeline here." He tapped the map. "It's rough terrain, but it'll keep us out of open ground and away from the likely patrol lanes."

Reilly leaned in, squinting at the contour lines. "That's going to be steep country, Lancer Six. If we hit contact, there's not much room to maneuver."

"That's why we establish a rally point," Jack replied. He circled a spot deeper in the hills. "Here. If we get split up, we regroup here at 1400. Anyone who doesn't make it by then moves alone back toward the fallback grid. No heroics, understood?"

A chorus of nods followed.

Davis flipped open his notebook. "What about comms?"

"Primary call sign for HQ is Iron Gate," Jack said. "Friendly lines will answer to Cedar Base. Our squad call sign is Lancer, I'm Lancer Six. Davis, you're Lancer Two-One on the radio. Keep transmissions short and clean—no names, no chatter."

"Roger that," Davis said, jotting it down.

Kim leaned forward, his expression focused. "Targets. What exactly do we need eyes on?"

Jack's pencil tapped across the map. "Enemy HQ, troop concentrations, any armor, artillery positions, supply depots, mortar pits, comms lines. If they're massing for an offensive, higher needs the full picture. Details matter—numbers, direction, even the condition of their equipment. If you see extra fuel drums stacked near an artillery piece, I want it noted."

Hill snorted softly. "So basically, if it moves, breathes, or makes noise, we write it down."

"Exactly," Jack said without humor. "But we stay invisible. No contact unless there's no other choice. The last mission worked because we stayed ghosts. Same rules this time."

The tent fell quiet for a moment, the team studying the map, committing routes and rally points to memory. Jack finally looked up, his eyes sharp but steady.

"We're not walking into Normandy again. We're sharper now. Smarter. We've been tested. This mission isn't about glory—it's about giving the boys on the line the information they need to survive. That's what matters."

Kim nodded firmly. "Understood, Lancer Six."

Reilly cracked his knuckles. "Let's do it, then."

Jack folded the map carefully and tucked it into its oil cloth sleeve. "Get some rack time. Tomorrow, we will prepare, check gear, and go over hand signals. On the 2nd, we move."

No one argued. The Misfits knew what was coming—and that every detail they had just memorized might be the difference between coming back together, or not at all.

The morning of January 1st came cold and gray, the kind of cold that crept through the seams of the tent and cut down into the bone. The Misfits had dragged their gear into a tight circle, each man surrounded by neat piles of ammunition, rations, and web gear. The air was filled with the soft metallic clicks of bolts being checked, magazines tapped, straps tightened.

Reilly ran a cloth down the barrel of his M1A1A1, muttering, "Clean rifle, clean kill." Hill was adjusting the straps on his pack, making sure nothing rattled. Kim sat cross-legged, working silently with a bundle of tripwires and a small pouch of grenades, his face calm, almost meditative. Davis crouched with his radio set cracked open, swapping out batteries and checking antenna connections, his lips moving as he rehearsed call signs under his breath.

Jack moved among them, checking each man's kit, asking a question here and there but mostly letting them work. His own M1A1 Carbine lay across his knees, bolt locked back, the magazine springs stretched and tested. Every so often, he would glance up, watching his men, pride settling in his chest. They weren't the same half-formed crew that had stumbled into Korea. They were sharper, tighter becoming exactly what he needed them to be.

When the moment allowed, Jack leaned back against his pack, letting the quiet rhythm of preparation fill the space. His mind, unbidden, slipped away from the cold tent and the weight of maps and missions. Patricia's face came to him her smile, the warmth of her hand on his arm, the way her eyes carried both strength and worry when she looked at him. She was half a world away, tending to the wounded men, fighting her own war in her own way.

For just a moment, Jack let himself wish he could be there, sitting beside her, sharing something as simple as coffee in the hospital ward instead of preparing to crawl through frozen hills hunting artillery. The thought tightened something in his chest, but he pushed it down, burying it the way he always did before a mission.

"Lancer Six," Davis said, snapping Jack back to the present. "Comms checks good. Call signs memorized. Radio's solid."

Jack gave a short nod. "Good work."

Hill held up a fist, each finger ticking off one of their rehearsed signals. "Advance, hold, enemy left, enemy right, fall back. All still the same, yeah?"

"Same as before," Jack confirmed.

Reilly snapped his ammo pouch shut with a grin. "Then we're ready. Just say the word."

Jack stood, slinging the carbine across his chest, and looked around at his men. His men. "Tomorrow, before dawn, we will step off. One more night of rest. Make it count."

The Misfits nodded and went back to their gear, but Jack knew none of them would sleep easily. Neither would he.

The world was still black when Jack opened his eyes. The tent was cold and quiet, the only sound the slow breathing of his men. For a moment, he lay still, listening to the faint hum of the wind outside, the muffled movements of a camp not yet awake. Then he eased up, careful not to rattle the canvas, and pulled on his gear.

One by one, the Misfits stirred. Reilly muttered something under his breath, tugging on his boots. Hill buckled his web gear with quick, practiced hands. Kim said nothing, his expression carved from stone as he slung his pack. Davis, the last to rise, cradled the radio set like it was a newborn, checking the antenna with a small click.

No one spoke more than necessary. They didn't need to. The weight of what lay ahead pressed down heavier than words. Jack checked his M1A1 Carbine one last time, sliding the bolt forward with a soft metallic snap. It felt solid in his hands, an anchor against the uncertainty waiting beyond the wire.

When they stepped out of the tent, the air hit them sharp and raw. Frost clung to the ground, crunching faintly under boots, and their breaths hung in pale clouds. The camp was dark save for a few guarded lamps, and the distant murmur of artillery thunder rolled over the hills like a reminder of why they were here.

Jack gathered them just outside the perimeter. He crouched low, voice barely above a whisper.

"Final check. Everyone good on ammo? Gear tight?"

A round of quiet nods answered.

"Route stays as planned. No contact unless forced. Davis, you keep chatter tight. We move silent, we move quick, we don't leave footprints behind."

The men leaned in; eyes fixed on him in the dim glow of the fading lantern light. Jack let his gaze settle on each of them, holding it for just a second longer than usual. They knew what it meant.

"Alright," Jack said, standing. "Misfits, move out."

The wire fence loomed ahead, shadowed against the faintest hint of gray in the eastern sky. Jack led the way, crouched low, breath steady, carbine ready.

Beyond the line, the hills waited—dark, cold, and filled with the unknown. The kind of place where men vanished, swallowed up without a sound.

But Jack Mercer and his Misfits moved like ghosts, the Misfits crouched low as they approached the wire. The last line of defense was little more than a stretch of coiled steel and a shallow trench, but it marked the end of safety. Beyond it, the hills stretched into blackness, an enemy land thick with shadows and listening ears.

Jack signaled for Hill and Kim to ease up the strands of wire with their gloved hands. The coils flexed with a faint metallic groan, just enough to let the men slip through one by one. Jack went first, the M1A1 Carbine snug against his chest, the frost crunching faintly under his knees as he crawled.

As he reached the far side, he froze. Voices drifted close—two guards on the inside of the wire, silhouettes shifting in the dim glow of a cigarette ember.

"…glad I'm not going out there," one of them muttered, his breath clouding in the cold air. "Those hills'll eat a man alive. Better them than me."

The other gave a low chuckle. "Ain't no pay worth walking into that."

Jack's jaw tightened, but he stayed still, holding up a clenched fist to stop the Misfits from moving. One by one, the guards' voices faded as they strolled down the line, the glow of the cigarette shrinking into darkness.

Jack let out the breath he'd been holding. He motioned the team through, and soon all six men were crouched in the dead ground beyond the wire. The camp, with its warmth and guarded safety, was behind them now.

He leaned close, voice barely more than air. "Alright. From here on—silent. Eyes up. Stay tight."

The Misfits nodded, their faces hard in the pale frost light. Then, like shadows, they turned toward the hills and began their long, silent march into enemy country.

Snow drifted over the ridgeline, muting the world into white and shadow. Jack Mercer crouched in the lee of a jagged rock, the wind clawing at the frost riming his eyelashes. The cold bit deep—sharp, metallic—but he barely noticed. Nights like this were a gift. Snow and darkness wrapped them in silence, cloaking the Rangers in a way no camouflage ever could.

"Lancer Six to Lancer Two-One. Authenticate – Fortitude-Gamma." Jack's voice was low, controlled, the throat mic catching the whisper clean.

The SCR-300 on his back hummed before the clipped reply came:
"Lancer Six, authenticated with Victor-Echo. Over."

Jack shifted his M1A1A1 Paratrooper carbine, the worn walnut grip dark against the snow-white over-smock. The folding stock creaked faintly, the same creak he'd heard in Sicily, Normandy, the Ardennes. It was a familiar voice in a foreign war.

His eyes moved across the fire team-five shadows in white, faces blackened, eyes sharp. Davis on point, Garand steady, wire cutters at his beLT Hill's Thompson wrapped in burlap to silence its rattle. Kim crouched low with the BAR, muzzle shrouded. Reilly moved like a ghost, medical satchel tight against his side.

They'd been belly-down for twenty minutes, letting the night settle. Ahead, the trench was a faint smudge, eighty yards away. A flicker of lantern light. Low wire. Sandbags. Chinese voices drifting through the snow.

Jack tapped Davis's shoulder, pointed. Davis slid forward, patient as a glacier. Wire cutters whispered—cut, twist, spread. The gap yawned open. Jack slipped through first, carbine leading.

At the trench lip, the smell hit him—coal smoke, sweat, unwashed wool. Jack gave the hand signal: stand-to. Hill and Kim split left and right. Davis dropped down. Jack followed, boots crunching softly.

The sentry barely turned before Jack's Fairbairn–Sykes needle-pointed dagger, drove up under the jaw, breath freezing mid-gasp. Jack eased him down without a sound.

From the far end came a grunt—Davis wrestling another man. Jack was there in seconds, binding wrists with waxed cords. The prisoner's breath fogged in frantic bursts, eyes wide above the gag.

"Lancer Six to Lancer Two-One," Jack whispered. "Package secure. Execute Icefall. Over."
"Lancer Six. Icefall acknowledged. Out."

They moved in silence. Snow fell harder, erasing their trail. Halfway to the ridgeline, a flare cracked overhead, magnesium light turning night into day. Shouts rose behind them.

Kim dropped to one knee, BAR chattering in sharp bursts, white phosphorus grenades hissing back into the trench. Hill fired his Thompson, muzzle flashes swallowed by the storm.

Jack shoved the prisoner forward. "Move, move, move!"

Bullets hissed past as they scrambled uphill, lungs burning. At the crest Jack keyed the mic: "Lancer Six to Lancer Two-One. Tick Five. Zero casualties. One package. Over."
"Lancer Six. Copy Tick Five. Proceed to Fallback Ridge. Out."

Only on the far side did Jack breathe. The prisoner stumbled, Davis yanking him upright. The storm smothered the chase.

Jack's glove brushed the scarred wood of his carbine. He had carried it through towns whose names these men had never heard, against enemies long dead. He carried it still—because some things, like the man himself, refused to age.

The war had changed. The cold had teeth. But for Jack Mercer, a fight was still a fight. And tonight, the Rangers won theirs.

The ridgeline fell behind them, the snowstorm swallowing every trace of the fight. For another half hour they pushed through the drifts, the prisoner stumbling between Davis and Kim, wrists bound tight, gag frozen stiff. The man's boots dragged furrows in the powder, but no one slowed.

Finally, the faint glow of lanterns bled through the storm—friendly lines. The barbed wire rose ahead, faint and jagged in the darkness. A challenge call rang out, sharp and wary.

"Thunder!" a voice barked.

Jack straightened, breath heaving from the climb. "Flash," he replied.

The Rangers slipped through the wire, MPs rushing forward to take the prisoner off Davis's hands. The man sagged in relief, but Jack felt no sympathy. He had too much to report.

Inside the command tent, heat and lamplight wrapped around them, the sudden warmth almost dizzying. CPT Walker stood over a map table; field phone pressed to his ear. He glanced up, eyes narrowing, then waved them over.

Jack pulled the gloves from his hands and began laying it out—every grid, every trench, every mortar pit and artillery position they had marked. He sketched arcs with a gloved finger on the map, his voice clipped, precise, the words tumbling out as the others added details: ranges, sight lines, estimated numbers. Davis described bunkers, Hill pointed out where the ammo dumps lay, Kim noted machine gun nests.

CPT Walker absorbed it all, jaw tight, then snapped the phone back to his ear.

"Higher, this is Blackjack Six. I have confirmed grids from recon element. Priority fire missions—standby for target data."

He relayed the numbers Jack rattled off, his voice sharp, leaving no room for misunderstanding. Outside, artillery batteries stirred awake, steel on steel clattering in the cold.

When he hung up, Walker finally looked at them, his expression iron, but his words carrying weight. "Outstanding job. You did more with six men than a company could've managed. The damage you just set in motion…" He shook his head. "You'll see the results before dawn."

He exhaled, tension easing only slightly. "Go get hot chow. Rack out. You've earned it. Be back here at 2000. I've got another mission for you."

Jack gave a single nod. The Misfits filed out into the snow, exhausted but sharp-eyed, the adrenaline still humming in their blood. Behind them, the map table glowed under lamplight, coordinates still etched across it—lines of destruction waiting to be written in fire.

Chapter 15 Eyes of Fire

The small mess tent was quiet except for the soft scrape of metal on canvas and the low hiss of the lantern swaying overhead. The air smelled of gun oil, coffee, and damp wool—an old soldier's perfume. The Misfits moved slowly, each man lost in thought, cleaning weapons that were already spotless.

They'd finished their debrief with CPT Walker an hour ago. Another mission briefing was scheduled for 2000. Until then, they waited—resting, if you could call it that.

Jack sat on an ammo crate, his M1A1A1 Paratrooper Carbine laid out across a field cloth on his knees. The folding stock gleamed faintly in the dim light, the walnut grip worn smoothly from years of use. Each motion—clean, oil, inspect, reassemble—was muscle memory, a ritual that kept his mind from wandering too far.

Across from him, Kim sat cross-legged, the BAR resting across his lap. He wasn't really cleaning anymore, just running the same rag across the receiver repeatedly. "You think he made it?" he asked quietly.

Jack didn't look up. "Who?"

"The sentry. The one in the trench."

Jack's hands paused, then resumed their work. "No. He didn't."

Silence filled the tent, heavy and familiar. The mission had gone perfectly—no mistakes, no casualties—but that didn't mean it hadn't cost something. Success always did.

Hill finally broke the stillness. "Walker said HQ hit those grids less than an hour after we got back. Whole sectors gone. Artillery flattened it."

Reilly gave a slow exhale. "We're good at what we do. Doesn't make it easier."

Davis, sitting near the tent flap, flicked a cigarette between his fingers without lighting it. "If we hadn't done it, more of our guys would be dead right now. That's the trade."

Jack tightened a screw on the carbine's receiver, eyes focused. "That's the job. We do what we have to so others don't have to."

No one replied. Outside, the cold wind rattled the canvas and whispered through the seams. The war never slept, even when they did.

After a long pause, Kim said, "You think this next one's gonna be worse?"

Jack gave a small, tired smirk as he locked the folding stock into place with a soft click. "They usually are."

The sound of metal meeting metal echoed softly. Jack checked the chamber, then slid the carbine's sling over his shoulder. "Get some rest," he said, voice calm but firm. "We'll need it."

He pushed through the tent flap, stepping into the pale daylight.

The cold hit him instantly, crisp and sharp. The camp stretched out under a bleached winter sky, the sun weak behind a layer of thin clouds. Snow crunched beneath his boots as he walked slowly toward the perimeter, past rows of tents and stacked ammo crates half-buried in frost.

A few men moved between tents, shoulders hunched, faces drawn and gray. Somewhere down the line, a cook shouted over a smoking stove, the smell of burned coffee drifting on the wind.

Jack stopped near the edge of camp, where the last line of sandbags met the open hills beyond. He rested a gloved hand on the cold metal of his carbine's stock, eyes scanning the ridgelines. Somewhere out there, the enemy was digging in again. Somewhere out there, another mission waited.

For a moment, his thoughts drifted— Patricia's face, her last letter, the warmth of a world that felt like a lifetime ago. He wondered if she'd still recognize him now.

He drew a slow breath, the air freezing in his lungs. The wind carried the faint sound of a distant engine, maybe a jeep or a supply truck. It all felt strangely normal for a war zone.

Jack adjusted his sling, turned back toward camp, and started walking. The snow squeaked under his boots, each step steady, deliberate.

2000 hours wasn't far off. Another briefing. Another set of coordinates. Another fight waiting to be fought.

And until then, just time enough to think too much.

Jack's boots crunched quietly over the frost-hardened ground as he approached the Company Command Post, a low sandbagged bunker tucked against the slope. The pale daylight highlighted the irregular mounds of earth and canvas flaps, but inside, voices carried urgency and tension.

He paused at the entrance, gloved hand resting lightly on the flap covering the doorway. Pushed open, it revealed a dim interior, the weak sunlight filtering through sandbags and casting sharp shadows over radios, maps, and scattered field gear.

Soldiers hunched over their stations, headsets clamped tight, shouting terse updates into the microphones. Jack crouched near the far wall, listening without being seen.

The radio crackled with hurried voices:

"…Company on the line, grid seven-three, heavy contact, push South—repeat, falling back…"

"…Artillery support needed…mortars on grid five-nine…South sector compromised…"

"…Company HQ, the Chinese are advancing. South is collapsing. Units can't hold—repositioning all reserves…"

Jack's jaw tightened. The words sank like lead. The Chinese were pressing hard, driving the Army South, and the company was on the brink of being overrun.

He stayed low, eyes scanning the sandbag walls and soldiers at the radios, noting tension in their shoulders and white-knuckled grips on headsets. Each crackle of static and clipped report painted the same grim picture—chaos creeping toward friendly lines.

Jack exhaled slowly, his breath fogging in the cold bunker. There was no panic in his stance, only sharp focus. Every mission he'd ever run had led to this: information, timing, and action.

He pushed back thoughts of Normandy, Korea, and Sicily. This was now. And if the Army was pushed south, someone had to go in and stop the advance.

Stepping back toward the flap, Jack let the voices fill the bunker as he moved. Outside, the pale sunlight glinted off snow-covered hills, but inside, the radio traffic left a bitter taste. The enemy was on the move—and the Misfits were going to be called into the heart of it.

Jack strode into the sleeping tents, alert and tense, his M1A1A1 Paratrooper Carbine slung over his shoulder. The first light of day filtered through the canvas, pale and cold. He moved among the Misfits, who were still curled in their bedrolls, half-asleep and reluctant to rise.

"SSG Mercer," Hill mumbled, eyes barely open.

"No time," Jack said, crouching beside him. "Get up. Chinese forces are pushing the Army south. We've got a serious situation, and we're going to be called in."

Kim stirred, sitting up and rubbing his eyes, BAR across his lap. "Already? Damn it…"

Reilly yawned, swinging his legs out of the bedroll. "Well, that sounds like trouble. What's the scoop, SSG Mercer?"

Davis was already upright, checking the radio, face tight with focus. Jack crouched in the center of the circle, scanning each man. "Command reports the South sector is under heavy pressure. We need to be ready—recon, support, extraction—whatever they throw at us, we respond. No hesitation."

Hill rubbed his jaw. "Sounds like hells on our doorstep."

"Exactly," Jack said, voice low and firm. "Gear up, check weapons, ammo, radios—everything. Nothing left to chance. We move as one unit, hit as one unit, and cover each other."

Reilly grinned faintly, loading his Garand. "Copy that, SSG Mercer. Let's get to it."

By the time the sun had risen fully, the Misfits were ready. Web gear strapped tight, boots laced, weapons cleaned and loaded. Davis double-checked the radio frequencies, Kim and Hill carried extra ammo and demolition packs, and Reilly gave his Garand one last inspection.

Jack rose last, slinging his carbine over his shoulder, eyes scanning the team. "We move out. Command Post first. We wait for the call, then step into the teeth of this fight."

They filed out of the tents in single file, the cold daylight biting at exposed skin. The crunch of frost underfoot marked their cautious march back to the sandbagged bunker. Inside, the low hum of activity and radio chatter greeted them.

Jack signaled the team to huddle near the far wall. "Everyone stays sharp. Eyes and ears open. We move the moment command calls. Speed, silence, precision—that's what keeps us alive."

The Misfits nodded, tension and readiness coiled tight in their bodies. Outside, the pale morning light glazed the hills. Inside the bunker, the radios crackled with reports. SSG Mercer and his team were awake, alert, and ready for whatever the day would throw at them.

Jack stepped into the sandbagged Command Post, the dim light casting long shadows across the walls lined with maps and radio equipment. The smell of coffee and the faint metallic tang of gun oil filled the air. CPT Walker looked up from a field map spread across a crate, his expression sharp.

"Jack," Walker said without preamble. "The push has been stopped at the 38th Parallel. Chinese forces held there, but it's been costly. HQ wants to reinforce the front. You'll be moving ten klicks north, right behind the lead elements."

Jack nodded, listening intently. "Understood, sir. What's the timeline?"

"Trucks will be here in thirty minutes to move the company north," Walker replied, tapping the map to emphasize the route. "I want your team ready to move with the HQ section. You'll only report to me—keep the Misfits close, but out of the thick of the line unless I call for you."

Jack straightened, voice steady. "Roger that, CPT Walker. The Misfits are just outside, fully geared and ready for anything you need."

Walker's eyes softened slightly, a rare flash of approval. "Good. I want your team to be sharp, quiet, and disciplined. No heroics—just do what needs to be done. You move with the HQ section, maintain communications, and wait for my orders."

Jack gave a crisp nod. "Yes, sir. I'll make sure they're ready to step on a moment's notice."

With that, Jack turned and left the Command Post, moving out into the cold daylight where the Misfits waited, weapons checked, packs secured, and eyes alert. The push north was coming, and he intended to make sure his team was prepared for whatever lay ahead.

The trucks rattled over frozen dirt roads, kicking up clouds of dust that quickly froze in the bitter morning air. Ahead, the terrain opened into rolling hills, dotted with foxholes, sandbags, and the skeletal frames of abandoned farmhouses. The distant sounds of artillery and small-arms fire were a constant reminder: the front was alive.

Jack's eyes swept the ridgelines. Smoke curls drifted upward from several points along the horizon—tracer fire lighting up the pale gray sky. The convoy slowed as they approached the forward position, a small cluster of sandbagged bunkers and makeshift observation posts. Soldiers moved methodically, reinforcing defenses and manning machine guns.

"Here we are," Jack said, voice low as he jumped down from the truck, M1A1A1 Carbine ready in his hands. The Misfits followed, landing quietly, their boots crunching softly in the frost.

CPT Walker was waiting at the command post, pacing in front of a large field map pinned to the sandbags. He looked up as Jack approached. "Good timing, Mercer. The company is dug in, defensive positions reinforced. You'll be with the HQ section, monitoring movements, marking enemy positions, and ready to insert anywhere we need you."

Jack gave a short nod. "Understood, sir. The team is ready."

Hill scanned the nearby tree line. "Movement over there—looks like patrols, maybe recon elements."

Kim dropped to one knee, BAR resting on his pack. "Range estimate—about 700 meters. Lightly armed. No armor, just infantry."

Davis keyed his radio, calling in coordinates back to HQ. "Lancer Two-One, reports sent. Keeping eyes on all enemy activity, sir."

Reilly checked the bindings on his pack, eyes darting to a distant hill. "Nothing too heavy, but they're spread out. Looks like they're probing for weak spots. Could get ugly if they press harder."

Jack crouched beside the map, scanning the terrain with sharp eyes. "We set observation points along the northern ridge. Davis, keep comms active. Hill and Kim, I want you to check flank approaches and ridgelines. Reilly, you're with me—secondary observation, quick-response."

The Misfits moved into position, methodical and silent. Frost rimed their eyelashes and brows, but the cold didn't slow them. Every ridge, every ditch, every shadow could hide a threat.

Jack's gaze lingered on the horizon. The front had shifted, the enemy had pushed south, and now they were here—ten klicks north of where the line had held. The Misfits' next hours would test them in ways that Normandy and the frozen hills behind them hadn't.

He inhaled sharply, tightening the strap on his carbine. "Eyes sharp. Stay quiet. And remember… we're the first to see it, the first to know it. The rest of the company counts on that."

A distant volley cracked the morning silence. The front wasn't calm; it had never been. But SSG Mercer and the Misfits were ready, moving into position as the silent shadow over the unfolding battle.

CPT Walker barked the order with a clipped efficiency that left no room for argument. "SSG Mercer, take the Misfits and Forward Observer LT

White up to that high ridge—roughly eight hundred meters west. You'll spot for fire on the enemy concentrations to the north. Move fast, move quietly, and keep comms clean. We want those positions suppressed before they can mass again."

SSG Mercer acknowledged with a curt, "Yes, sir." He turned to his men. "Pack light. Move as a single file. LT White will call the shots — we're his eyes, his security, and his pencil. Everyone ready?" The quick, tired nods around him were all the confirmation he needed.

LT White was a narrow-shouldered lieutenant with a forward observer vest and a map board strapped to his chest. He moved with the brisk, precise air of a man lived by coordinates and timing. "I'll need a clear line of sight to the northern approaches," he said, tapping his compass. "If we get that ridge, I can bracket their positions and start an adjustment. Keep me alive long enough to get rounds on target." His voice was flat, professional, no bravado, only the hard fact of the work.

They slipped away from the command post and across the frozen ground in a tight, measured flow. Snow bites kept sound at a minimum, swallowed by powder, breath low and even. The ridge was a dark seam on the horizon, a jagged silhouette of rock and scrub. As they moved, SSG Mercer scanned constantly, noting gullies that could hide an approaching patrol and low saddles that might offer better cover. Kim and Hill took the flanks; Reilly and Davis scouted slightly ahead and behind. The team's movements were a choreography of practiced caution.

When they topped the ridge, wind hit them full and raw. From that height the north lay open: fields, broken ground, and a scattering of enemy encampment smoke curling from cookfires, the glint of tarps, and the dark shapes of gun emplacements. LT White dropped to one knee, snapping his binoculars up and then settling om his map board. He called coordinates into his notebook, running quick compass checks and marking reference points in a tight, exact hand.

"Range to primary cluster: 2,100 meters. Secondary emplacements at 1,800 and 2,350," LT White said, voice low but steady. He keyed his handset to the battery nets and spoke the fire mission in the same clipped numbers and modifiers Jack had heard a hundred times before. "Falcon Fire Control, this is White Actual. Adjust fire for grid—bearing—elevation…" He read the data off, and the radio swallowed the call as artillery crews at the batteries began their ritual answers.

SSG Mercer and the Misfits arranged themselves for security. Kim set up the BAR across a low rise; barrel shrouded against the snow glare.

Hill found a shallow depression to cover the western approach. Reilly kept an eye on the valley; Davis relayed periodic updates to CP on a secondary channel. Every man felt the familiar, taut mix of dread and resolve as the first calls went out.

Minutes later the first salvo arced in—distant thunder that rolled into the valley and then a muffled, violent impact. Dirt and snow flew where the shells struck; a plume of gray smoke marked the first adjustments. LT White murmured a correction, and the next round fell closer, this time cutting through the enemy ring like a scythe. He barked, "Shift left twenty, add two! Danger close—hold fire for civilians," and SSG Mercer felt every syllable like a pivot point: precise, ruthless, necessary.

They watched as a rhythm formed: spot, call, impact, adjust. Each explosion tightened the enemy's options and bought time for the front below. SSG Mercer checked his men—faces set, fingers whitened on triggers—and let himself register, briefly and dangerously, a small, cold satisfaction. The Misfits had brought the eyes; White would bring the fire. For now, their job was to keep those eyes open.

Part 4: Orders and Echoes

Chapter 16 Midnight Orders

Snow drifted lightly across the ridge, the flakes fine as dust and just as relentless. For thirty-six hours, SSG Mercer and the Misfits had held their position alongside LT White, watching the valley below through binoculars and scopes while the artillery did its brutal work.

The once-still landscape was now a patchwork of blackened craters and churned snow. Smoke still curled from shattered tree lines, and the faint echo of distant guns rolled over the hills like a heartbeat that wouldn't stop.

Jack crouched near a rocky outcrop, his M1A1A1 Paratrooper Carbine resting across his knees. His face was drawn, eyes sunken with fatigue, but his focus was as sharp as ever. He'd seen long watches before—Normandy, the Ardennes, Holland—but the Korean cold was something else entirely. It gnawed at the bones and hollowed men out.

LT White stood a few paces away, muttering into the radio handset, his voice low but confident. "Fire Control, this is White Actual—final adjustment complete. Rounds on target, effects observed. We're disengaging and heading back to Command Post, over."

Static crackled. Then came the reply, faint and tinny: "Copy that, White Actual. Good work. Safe travel back."

White exhaled, rubbing the frost from his gloves. "That's it, Sergeant. Missions wrapped. Let's move before we freeze solid."

Jack gave a curt nod and rose, stretching the stiffness out of his legs. "You heard the Lieutenant, Misfits—pack it up. We're heading back to the CP."

The men moved with practiced efficiency, rolling up blankets, checking weapons, slinging their rucks. Davis broke down the radio antenna while Reilly kicked snow over the small fire pit they'd used to heat rations. Hill checked the ridgeline one last time through his field glasses, then lowered them slowly.

"Looks quiet," Hill muttered. "Too quiet."

"Quiet's good," Kim said, slinging the BAR. "Means we live to do this again tomorrow."

Jack adjusted his harness, giving one final scan of the valley. "Let's move. Double interval, stay low. Last thing we need is to get caught with our backs to the ridge."

They began the descent, moving in staggered intervals down the frozen slope. The snow muffled every step, but the silence wasn't comforting—it was the kind of silence that pressed against the ears, waiting to be broken.

Jack glanced back once as they moved. The ridge where they'd spent the last thirty-six hours was already vanishing in the haze of smoke and falling snow. Another patch of cold earth they'd held another piece of the war that would fade into the white.

"Keep it tight," he called softly. "CPT Walker's waiting. Let's go home."

They had been moving for nearly twenty minutes, the ridge fading behind them, when the first sound cut through the stillness—a single, distant crack. Not artillery. Not thunder. A rifle shot.

Jack froze mid-step; one hand lifted in signal. The team dropped instantly, spreading out into shallow depressions along the hillside. Hill slid into cover beside a half-buried boulder, scanning through his field glasses.

"Talk to me," Jack murmured.

Hill adjusted the focus. "About four hundred meters—movement in the trees below the north slope. Small group. Six, maybe seven. Chinese scouts, looks like."

LT White crouched beside Jack, his face pale beneath the grime. "They must be moving to reoccupy the line we just shelled. If they spot us, they'll call in more."

Jack's eyes swept the terrain. The Misfits were exposed—thin cover, no high ground left to retreat to. "We can't fire," he said quietly. "Too close to friendly lines. We draw attention, we'll bring half the valley on us."

Kim shifted his BAR into position anyway, eyes cold. "Then we wait?"

Jack nodded. "We wait."

The next few minutes stretched like hours. The enemy patrol moved slowly through the trees, rifles slung, scanning the hillsides. Jack could hear their voices faintly on the wind—calm, unaware.

Then, just as suddenly as they had appeared, the scouts turned north, heading deeper into the valley toward the smoldering ruins of their camp.

Jack exhaled. "Let them go." He motioned for the team to move. "Stay low. Double-time. We're ghosts again."

The Misfits moved quickly, every step measured and deliberate. The snow muffled their passage, but tension hung heavily in the air. They were less than five hundred meters from the forward lines now, and the distant crackle of friendly machine guns told them they were nearing safety.

By the time they reached the outer perimeter, MPs were already waiting to wave them through the wire. Jack gave the all-clear signal, and the team slipped back into the safety of the company sector.

The familiar sight of the sandbagged command post came into view—smoke rising from chimneys, men moving with purpose, the organized chaos of a front-line HQ.

Jack glanced back at his team, each man worn, frost-covered, and silent but alive. "Good work, Misfits," he said quietly. "Now let's go report in."

They crossed the line together, heading for CPT Walker's bunker.

They crossed the wire and moved straight to the sandbagged command post, boots silent on the packed earth. Inside, the bunker smelled of hot coffee and burnt oil; men hunched over maps, radios hissed, and a chalkboard listed units and status updates in a tight, precise hand.

CPT Walker looked up as they entered, hands already on a pencil and a clean sheet of paper. The map on the table had new red marks where Jack's team had directed fire. He didn't waste pleasantries.

"Report," Walker said.

Jack stepped forward, shoulders squared despite the fatigue. "SSG Mercer, sir. We held the ridge with LT White for thirty-six hours. Observed enemy emplacements and supply movement north of that initial line. Callouts confirmed multiple artillery and mortar pits, plus forward supply dumps. We called in fire and adjusted until White reported effects observed. We disengaged and returned without casualties."

Walker's pencil scratched as he annotated the map. "Good. Specific numbers, locations, movement patterns. Don't gloss over anything."

Jack delivered it tight and clean: grid references, estimated crew sizes, the orientation of gun barrels, the number of trucks on the supply track, temporary bivouacs in the tree line. He described patrol patterns and timing—when scouts moved, how often runners passed between emplacements, where mortarmen sheltered. Hill and Kim filled in flank details; Davis handed over a time-stamped page of coordinates.

Walker studied the cluster of hashes on the map. "These positions here—secondary artillery and the mortar pits—were they being camouflaged or retired?"

"They were camouflaged and active," Kim answered. "Crews shuttled rounds in, hauled them out on sleds and by hand. This looked like sustained firing capability, not a temporary stand-up."

Jack added, "We also saw defensive works being reinforced—sandbag revetments, foxholes widening into firing positions. But the steady stream of fuel drums and extra trucks suggests more than just hunkering down. Could be resupply for a push or stockpiling to hold. We need deeper recon to determine intent."

Walker tapped the map, eyes hard. The air in the bunker seemed to tighten. Outside, the valley was still smoking from their strikes; inside, the shape of the next move had to be decided.

"All right," Walker said finally. "HQ wants follow-up recon in depth. I want eyes inside that support belt—close enough to count vehicles, observe refit patterns, and, if possible, snatch anything useful: documents, radios, prisoners. Highest priority: determine intent—defense or offensive massing."

He rolled the map to reveal a corridor of gullies and ravines leading toward the enemy concentrations. "SSG Mercer, you and the Misfits will go. Approach from the west, use the gullies—minimize exposure in the valley. LT White will be on a support hold; artillery will be on standby but will not fire unless you request or are compromised. Your staging point will be Fallback Ridge two klicks west of the objective. Davis will run comms—periodic, scheduled check-ins only. Move light, move fast."

Davis's fingers tapped the radio case. "Understood. We'll be disciplined."

Walker paused, then adjusted the timetable. "You'll move tonight. Not immediately—ten hours from now. Use the time to plan, rest, and prepare. I don't want you to run on fumes for a job that demands sharp eyes and steady hands. Eat, sleep, rehearse the approach and exfil routes. Be ready to step off at 0200. Full after-action required—every grid, every minute."

Jack nodded, relief and resolve both tightening his reply. "We'll be ready, sir. In and out. No unnecessary contact. We bring back proof— paper, radio, or prisoner—or nothing."

Walker gave the map one last hard look. "You have authority to call for suppression if you're compromised. HQ has approved the mission. Move light. Move clean. Bring them back."

Jack folded the inked page and slid it under his arm. The map felt heavier now, not from the paper but from the choices and risks inked on its surface. He turned toward his team, the cold air outside hitting them as they left the bunker.

Ten hours. Time to plan, to sleep, and to sharpen. The Misfits moved to their tents and set about the slow, precise work of preparation— packing, revising routes, rehearsing hand signals, and stealing a few hours' sleep while the world outside grew dim toward night.

Snow whispered against the tent's canvas, a thin, steady sound that filled the pauses between words. The lantern burned low, its flame a trembling bead of light that cast the Misfits' faces in amber and shadow.

Jack sat at the center of it all, the map unrolled before him—grids, contours, and pencil marks dense with meaning. For two hours, he had gone over every possible angle of the upcoming recon: where they'd stage, how they'd move, what signals to use if cut off. Every "what if" had been spoken aloud and solved, every escape route drawn and redrawn until the men knew them by memory alone.

"Patrol routes here and here," Jack said, tapping the paper. "If they're massing for an attack, we'll see supply movement along this track before dawn. If they're digging in, look for fixed positions—sandbags, wire, maybe gun pits. We keep eyes open, move quiet, and don't linger."

Davis nodded, jotting notes. "What about if we run into one of their scout lines early?"

"Break contact," Jack replied. "No firefights. We're not here to win one—we're here to know what's coming."

Hill exhaled, rubbing his hands together for warmth. "Feels like we're walking straight into the dragon's mouth again."

Jack gave a faint smile. "That's where the answers usually are."

After another half-hour of planning and dry-run discussion, the tent fell into silence. The maps were rolled, pencils capped, and weapons checked one last time. Each man settled into his bedroll with that strange, restless quiet that came before a mission—half exhaustion, half anticipation.

Jack lay back, eyes tracing the seams of the tent overhead. His mind should have been on the terrain, the routes, the radio calls—but instead, it drifted to Patricia.

He could almost smell her perfume beneath the faint tang of gun oil; soft and clean, like lilacs after rain. The memory of her lips, the warmth of her touch, lingered as real as the cold ground beneath him. Her voice echoed softly in his mind— gentle, certain, telling him to come back safe.

For a moment, the war faded. There was no Korea, no frost, no maps—just her. The world he fought to return to.

Jack closed his eyes, the ache of memory and duty mingling in his chest until sleep finally crept in, silent and heavy.

Outside, the wind sighed through the camp, carrying the low rumble of artillery far to the north—the reminder that when dawn came, they'd walk straight back into the storm.

The lantern had long burned out when Jack opened his eyes. The tent was dark and still, the cold seeping through the canvas and into his bones. He checked his watch—0000 hours.

It was time.

He dressed quietly, slipping on his parka and boots, the chill biting at his fingers. His men were still asleep, breaths steady and even. Before waking them, Jack stepped out into the frozen camp.

The night air hit him like a slap—sharp, clean, and full of the low hum of generators and distant voices. A thin mist hung over the ground, glowing faintly under the hooded lamps strung between bunkers. Jack made his way toward the command post, boots crunching softly in the snow.

The sandbagged bunker was alive with quiet urgency. Inside, the air was thick with the smell of coffee and damp wool. A sergeant hunched over a radio, headphones pressed tight, while a pair of officers traded clipped updates over a map covered in grease-pencil marks.

CPT Walker looked up from the table as Jack entered. "Mercer. You're up early."

"Wanted to check in before we step off, sir," Jack said, his voice low but steady. "Anything new from higher?"

Walker rubbed his eyes and gestured toward the map. "Enemy movement north of the river. Scouts report heavy traffic along the supply roads— trucks, pack animals, fuel drums. Could be resupply, could be staging. No one knows for sure yet. That's why I need your team to confirm it."

Jack studied the marked positions, the red lines creeping south like veins. "Any change in our sector?"

"None yet. The line's holding, but it's tense. Chinese activity's been probing near the ridges— feels like they're mapping us as much as we're mapping them."

Jack nodded slowly. "Understood. We'll bring back answers."

Walker's expression softened just slightly. "Step off at 0130. Keep it quiet, Mercer. I'll be monitoring your check-ins. If anything changes, I'll relay through Davis."

"Yes, sir."

Jack saluted lightly, then turned and stepped back into the freezing dark.

The camp was still as he approached the Misfits' tent again. Inside, the lantern flickered low, casting long shadows across their sleeping forms.

He crouched and placed a hand on Davis's shoulder.

"Up," he whispered. "Midnight. Time to move."

Davis blinked, then nodded and sat up, already reaching for his radio. Jack moved to each man in turn—Hill, Kim, Reilly—waking them with a quiet word and a steady hand.

Within minutes, the tent filled with the soft, familiar rhythm of preparation: bolts checked, straps tightened, canteens muffled. Jack reassembled his M1A1A1 Paratrooper Carbine, the metal cold but familiar under his hands. He pulled back the bolt, checked the chamber, and slid it forward with a soft metallic snap.

At 0115, he stood. "Final checks. We move at 0130."

The Misfits rose without complaint, faces drawn but determined. Hill muttered, "Colder than hell out there."

"Colder means quieter," Kim replied, slinging his BAR.

Davis adjusted the radio, testing the frequency. "Cedar Base, this is Lancer Two-One, comms check—over."

Static, then: "Lancer Two-One, this is Cedar. Read you five by five. Good hunting."

"Roger, Cedar. Lancer out."

Jack pulled back the tent flap, wind howling against the canvas. The sky was pale with starlight, the hills to the north black silhouettes against the snow.

He looked over his men—his Misfits—and gave a short nod. "Same order as before. Stay low, stay silent. We've got a long walk and a longer night ahead."

"Roger that, SSG Mercer," Davis replied.

They slipped into the cold, moving toward the wire like ghosts. Behind them, the camp slept on. Ahead, the ridges waited—silent, frozen, and full of secrets.

Chapter 17 Eyes on the Valley

The night was black and brittle, the air sharp with frost. The Misfits moved like shadows through the broken ground beyond the wire, each step measured and deliberate. Snow whispered under their boots as they climbed, the pale starlight barely enough to mark the rise of the ridgeline ahead.

Jack led them, M1A1A1 Paratrooper Carbine tight against his chest, eyes scanning the slope. "Keep spacing tight," he whispered. "No lights, no noise. We stay in the dark until first light."

They angled uphill, cutting across frozen brush and outcroppings of rock. The climb was steep, the cold biting through their layers with every breath. Somewhere below, the front lines were little more than a murmur of distant engines and the faint crackle of radios. Ahead, nothing but the hill, the wind, and the weight of the mission.

The night thinned toward gray as dawn crept over the mountains, faint light brushing the ridgeline where Jack and the Misfits crouched. They had reached the crest just after 0500, breath steaming in the frozen air. Below them stretched the valley—wide, cold, and alive with the subtle stirrings of an army in motion.

Jack raised his binoculars and scanned the darkness softening into morning. The faint outlines of tents and smoke plumes told the story: the Chinese were dug in deep and busy, moving supplies, reinforcing positions, and laying down fresh communication wire.

He lowered the glasses and turned to the others. "We hold here," he said quietly. "Good field of view, cover from the ridge behind us. We'll find a hide and start logging movement."

Hill nodded and moved toward a deep drift near a boulder the size of a truck. He dropped to one knee, unstrapping his entrenching tool.

Jack watched, puzzled at first, as Hill began to carve into the drift—slow, deliberate strokes cutting a neat oval into the snowpack. "Hill," Jack called softly, "what are you doing?"

"Making us a home, Sarge," Hill said without looking up. "This drift's packed solid and the rock behind it'll block wind. I'm digging out a snow cave—big enough for three if we squeeze. We can rotate inside for rest and warmth. Candle'll heat it enough to keep frostbite off."

Jack crouched beside him, watching the smooth walls take shape. It wasn't just a hole—it was craftsmanship, the kind learned in cold nights from one winter to the next. "Good thinking," Jack said. "Make sure the opening faces the rocks so we're not sky lined. Kim, help him finish it out. Reilly, pack snow around the lip once it's hollow."

"Yes, SSG Mercer," Kim answered, dropping beside Hill to widen the tunnel. Snow flew in steady rhythm, dull against the frozen wind. Within fifteen minutes they'd carved a shallow dome into the drift, walls smooth, a narrow entry angled toward the rock outcrop.

"Looks good," Hill said, sliding back out. His face was flushed from the effort. "We can fit three at a time. Candle in the center'll take the edge off."

Jack gave an approving nod. "Rotate two hours in, two out. No light leak. Davis, you're up on the first watch with me."

The men settled into position, half in the cave, half prone behind rocks overlooking the valley. Davis extended the radio antenna beneath a low branch and listened to the faint static between scheduled silences. Jack keyed the radio mic. "Cedar Base, this is Lancer Six. Observation position established, ridge line west of grid three-four-one. Beginning surveillance of enemy valley positions. Over."

A faint reply crackled through. "Lancer Six, Cedar copies. Maintain radio silence unless critical. Good hunting. Out."

Jack exhaled and lowered the handset. "Alright, Misfits," he said quietly. "We've got our nest. Let's get eyes on everything that moves. Walker needs the truth—and we're the ones who'll find it."

Jack lifted his binoculars again, scanning the gray valley below.

Movement.

"Contact," he murmured. "North end of the valley—convoy, five trucks, maybe six. It looks like fuel or ammunition. Mark it, Davis."

"Copy," Davis said, pencil scratching against paper.

To the east, another group of figures trudged between half-buried shelters, stacking crates beside a mortar pit. Kim adjusted the scope on his rifle and tracked them, whispering counts under his breath. "Two dozen there. All working fast."

"Looks like reinforcement, not retreat," Reilly said quietly.

Jack nodded, eyes narrowing. "Agreed. They're massing, and digging in. CPT Walker needs to see this."

The morning light grew brighter, painting the valley in pale blue. From their high perch, the

Misfits could see nearly everything—the thin trails of smoke, the silhouettes of men moving between tents, the faint gleam of vehicle metal beneath camouflage nets.

They took turns recording every sighting, every pattern, every truck that came or went. Between shifts, Hill and Kim crawled into the snow cave, lighting a single candle to chase off the bite of the cold. The small glow flickered against packed white walls, and the faint warmth felt like a luxury.

Outside, Jack remained at his post, glassing the valley through the rising light. His breath hung in clouds around him, his mind already piecing together what the movement meant.

They weren't just holding this line—the Chinese were preparing for something big.

And the Misfits were the only ones close enough to see it happen.

The sun was still a weak smear behind the clouds when Davis leaned toward Jack, radio handset pressed against his ear. "Signal's clear, Sarge. Cedar Base is standing by."

Jack nodded, eyes never leaving the valley below. "Transmit it."

Davis keyed the handset. "Cedar Base, this is Lancer Two-One. Message traffic from Lancer Six, priority intel report. Over."

"Send it," came CPT Walker's clipped reply through the static.

Jack leaned close, voice calm but firm. "CPT Walker, this is SSG Mercer. We're in position overlooking the valley west of grid three-four-one. Observed multiple enemy movements—truck convoys, infantry work details, supply sleds. Movement is consistent and deliberate."

"Roger that," Walker said. "What's your read, Mercer? Offensive staging or defensive prep?"

Jack took a breath; binoculars still trained on the frozen valley. "Professional opinion, sir? They're digging in. Pattern of work lines, trench shaping, gun pits being reinforced. No signs of troop massing or pre-assault movement. They're building a defensive line—probably bracing for us to move north again."

There was a pause on the line. Then Walker's voice returned, lower, thoughtful. "Copy that, Lancer Six. Good work. Maintain observation and keep logging coordinates."

"Wilco, sir. Lancer Six out."

Jack nodded to Davis, who shut the radio down to standby. The team went back to their instruments—binoculars, scopes, and notepads. The morning dragged on in slow rhythm: mark, count, confirm. Even the cold seemed to fade under the focus of their work.

Then Kim stiffened beside him. "Jack… look."

Jack followed his gaze toward the eastern edge of the valley. Between the haze and snow, large shapes were emerging—dark, deliberate silhouettes dragged by tracked vehicles. At first glance they looked like trucks, but the long barrels gave them away.

Artillery. Big ones.

Jack's gut went cold. "Jesus," he muttered. "That's not defense—that's heavy guns setting up for range."

Davis was already reaching for the radio. Jack keyed his mic. "Cedar Base, this is Lancer Six, break—new movement. I say again, new movement. Enemy is positioning heavy artillery on the east side of the valley, grid three-four-five. Dug-in positions, multiple pieces. Estimate—six to eight guns, possibly one-twenty-two millimeter. Over."

The reply came fast. "Copy that, Lancer Six. Hold position. Can you confirm intent?"

"They're building firing pits, sir," Jack said. "Looks like they're setting up for sustained bombardment."

A moment's silence—then CPT Walker's tone changed, brisk and decisive. "Mercer, I need eyes and exact data. I'm sending a fire control element forward. Can you receive additional personnel?"

Jack's gaze swept their hide. The snow cave, the narrow ledge, the cramped line of sight. "Negative for a full section, sir. Position can only handle two more without compromising concealment."

"Understood," Walker replied. "Then I'll send LT White and myself. I'll bring the RTO when the line's stable."

Jack hesitated. "Sir, with respect, that's too many bodies up here. The terrain won't hide more than six of us without giving us away."

Walker's voice came back calm but firm. "I need White's eyes on the guns, and I need my own to verify targeting data. You'll send two of your men back to guide us in."

Jack exhaled, looking around at his team. "Understood, sir. Hill and Kim will move to meet you. They'll guide you in from the front line."

"Roger that," Walker said. "ETA?"

Jack checked his watch. "Give them ninety minutes, sir. They'll reach your position by then. Once you're ready to move, have them signal by lamp, low beam—one flash every five seconds."

"Copy all. Lancer Six, maintain surveillance until we arrive. Don't let those guns disappear. Blackjack Six out."

Jack lowered the handset and rubbed his gloved thumb across the carbine's stock. The wind was rising, scattering fine snow across the rocks. Down in the valley, the Chinese crews were still at work—heaving earth, aligning barrels, and stringing ammunition lines.

He turned to Hill and Kim. "Pack light, move fast. Get to CPT Walker's CP and bring him and LT White back here. No noise, no light. You'll guide them straight to this ridge. One and a half hours, understood?"

Hill adjusted his sling. "Understood, Sarge."

Kim nodded. "We'll get them here."

"Good," Jack said. "Now move."

The two men slipped out of the hide, disappearing into the whitening air. Jack watched them go, then turned back to his glass. The valley below was still moving—men, trucks, and now guns. The Chinese weren't just building a line.They were preparing to fire.

Hill and Kim moved fast but low, the wind cutting across their faces like wire. The sky had dulled to slate gray, snow drifting in tight eddies around their legs as they slipped down from the ridge. The world was quiet—too quiet. Only the whisper of boots through crusted snow and the faint creak of their gear marked their passage.

They navigated by memory and instinct, retracing the route they had taken hours earlier under darkness. The valley below was alive now—distant shouts in Chinese, the metallic clank of tools against frozen ground, the groan of engines straining to move the heavy guns Jack had reported.

Hill raised a fist and dropped to one knee, signaling haLT Ahead, faint silhouettes moved along the lower slope—four, maybe five men, their uniforms dark against the snow. The enemy was digging new trenches, reinforcing the lower valley.

Kim crouched beside him, breath frosting the air. "We go wide," he whispered. "Skirt left through the rocks."

Hill nodded. "Slow and low."

They slid through the boulder field, each step deliberate, avoiding the crunch of brittle ice. Twice they froze as the enemy voices grew louder, once when a flare hissed into the air and turned the snowfield to daylight. They pressed flat against the ground until the light faded, and darkness returned.

When they finally crested the last rise overlooking friendly lines, Hill could see the faint glow of lamps in the trenches below. Relief flickered briefly through his chest—but only briefly.

"Let's move," he murmured. "Walker's waiting."

They slipped through the wire under the watchful eyes of the sentries, giving the pre-arranged signal—two short clicks on a flashlight covered by a gloved hand. The guards waved them through without a word, and within minutes, Hill and Kim ducked into the sandbagged bunker that served as the company command post.

Inside, the air was heavy with tobacco and radio static. CPT Walker and LT White bent over a field map, their faces drawn tight with fatigue. Both men looked up as Hill and Kim entered.

"Report," Walker said.

Hill saluted quickly. "Sir, SSG Mercer sent us. He's confirmed Chinese artillery—six to eight heavy pieces being dug in east side of the valley. He requests you and LT White only. Position can't conceal more than six men total. He's maintaining observation until you arrive."

White straightened, already reaching for his binocular case. "He's sure about the calibers?"

Kim nodded. "Positive, sir. Long barrels, big crews, full ammo chains. Looks like they're building for sustained fire."

Walker exchanged a look with White, then gave a curt nod. "Grab your gear. We move in five minutes. Mercer's counting on us."

The wind had picked up, dragging thin veils of snow across the valley floor. Hill led the small column—Kim was close behind, with CPT Walker and LT White following a few paces back. Their breath plumed faintly in the dim light as they picked their way through the broken terrain, every sound muffled by the snow.

"Keep it slow and quiet," Hill whispered. "We cut through the low ground, then break for the ridge once we're clear of the valley floor."

Kim nodded, scanning ahead with sharp eyes. "Tracks everywhere," he murmured. "Chinese patrols have been working this area all morning."

They moved carefully through the low ground, weaving between clumps of frozen brush and shell-cratered depressions. Every few minutes they dropped flat as the sound of voices drifted through the cold air—enemy patrols moving along the slope just ahead.

Hill raised his fist and motioned them down. The four men pressed against the icy ground as a squad of Chinese soldiers trudged past less than thirty yards away, their silhouettes gray shapes in the swirling snow. Rifles slung, heads low against the wind, they spoke quietly to one another before vanishing into the haze.

Walker's gloved hand tightened around his rifle. He glanced at Hill, who held up two fingers— wait. Another patrol followed a minute later, moving in the opposite direction. For a long moment, the four of them stayed motionless, listening to the fading crunch of boots and the faint jingle of gear.

Only when the last sound had faded did Hill signal to move. They shifted direction, angling west, hugging the edge of a shallow gully that offered a narrow strip of cover.

LT White crouched beside him as they climbed. "How much farther to Mercer's position?"

Hill checked his compass and the slope of the terrain. "Another three hundred meters. Once we hit the base of that rocky ridge, we'll swing east and come in from behind their hide."

They were halfway up when Kim suddenly stopped, dropping to one knee. "Patrol—two o'clock," he hissed.

A pair of figures were moving along the base of the hill, slung rifles, scanning the ground for tracks. The lead soldier stopped and crouched, examining something in the snow.

Hill's heart thudded in his chest. He raised his carbine, finger resting lightly on the trigger guard—but the soldier stood, shook his head, and motioned for his partner to move on.

When the patrol disappeared over the rise, Hill exhaled slowly. "That was too close."

Walker gave a grim nod. "Welcome to Mercer's world."

They climbed again, moving faster now, the steep terrain and snow pulling at their legs. The higher they went, the quieter it became—no voices, no movement, only the sound of wind scraping over rock.

At last, they reached the upper slope where jagged boulders crowned the ridge. Hill raised a hand, signaling halt, then gave a short, low whistle—the prearranged signal.

A moment later, a faint answering whistle came from above.

Hill turned to Walker. "That's them, sir. We're clear."

The small group climbed the final stretch, pulling themselves up over the rocks until the narrow hollow came into view. Jack was crouched near the entrance of the snow cave, binoculars in one hand, his M1A1A1 resting across his knees. Davis looked up from the radio, relief flickering across his face as the newcomers appeared.

Walker and White crawled the last few feet into the hide, snow crunching under their boots.

Jack gave a curt nod. "Glad you made it, sir. You ran the gauntlet down there."

Walker dropped beside him, brushing frost from his shoulders. "We did. Now let's see what we're up against."

Jack handed him the binoculars and pointed toward the valley below, where smoke and movement marked the growing artillery positions. "There it is sir. Everything you asked for—and maybe more."

LT White laid out the sequence without fuss, the practiced economy of a man who'd watched too many rounds fall and learned how to make each one count. He tapped the map board and then the valley below, eyes steady.

"I want it done methodically," he said. "We'll use high-explosive for adjustment — single rounds, one at a time — until we bracket the primary gun pits. Once we have lethal effects and can mark the target we drop a smoke round to mark the impact area for the strike package. When the smoke is in place, the Air Force moves in for the bombs. After the bombs hit, we set fire for effect with the batteries — thirty minutes minimum. Flatten the revetments, disrupt their ammo chains, and make their positions untenable."

Walker listened; pencil poised over the map. "Single HE for adjustments first. No walking rounds and no hurry. We need accuracy more than volume until we're sure of the lay."

White nodded. "Exactly. Call for the first round, watch impact, adjust by meters — left, right, add, drop — whatever it takes. When the rounds are consistently landing on the emplacement, send a smoke to mark. That ensures the pilots can identify and won't mistake the wrong cluster. After the smoke, you'll get the strike. Bombs first to break the hardened pits, then the artillery will finish the suppression and finish any survivors."

Jack listened, fingers nipping at the rim of his glove. "What about civilians or friendly positions down there?" he asked. "We're close to our own lines in places."

Walker met his eyes. "Danger close concerns are on the table. We'll account for the grid offsets and push fire direction away from our troops. LT White, can you get adjustment brackets onto the guns without exceeding danger-close limits?"

White tapped his compass and calculator. "Yes. We'll use measured bracketing and keep initial rounds on the far side of the pits, away from our line. I'll call danger close if we must tighten the bracket, but I'll only do it when you authorize, sir."

Davis, hands already on the handset, keyed the net and spoke the data cleanly: target grids, range, bearing, requested rounds, and fire units. "Falcon Fire Control, this is White Actual. Request adjustment fire on grid 345-812. Initial fire: one HE, observer adjustment. Over."

Static, then the terse voice of fire control: "White Actual, copy. Battery one ready. Await your call for first round."

Jack and the Misfits settled into their roles—White as the voice and pencil, Walker to coordinate and authorize, Jack to keep eyes on the valley and the safety of the team. Kim and Hill tightened their positions to shield the observation slit; Reilly readied medical gear and warmers; Davis watched the net and kept time.

The first round left the tube as a distant mechanical cough. All of them watched the valley through optics and the naked eye, waiting for the telltale plume. The shell struck short of the emplacement. White murmured an adjustment, then called it in. Another round, and the plume of snow and dirt shifted closer. A third round cut through the frozen soil and struck in the revetment properly dirt and splintered timber flew skyward.

"Impact—on target," White said quietly. He sent the correction into the handset: "Adjust left twenty, add two." The battery complied. The next round landed squarely in the gun pit, sending a geyser of snow, wood, and black smoke.

"Mark with smoke," Walker ordered. "Make it visible to the air package."

White called for the smoke round. It arced slow and low and then blossomed in a thick white plume over the emplacement, a bright, furious beacon against the gray valley. Moments later a distant, deep roar announced the arrival of the strike package — aircraft running in from the north.

Jack felt the earth under his boots thrum as the first bombs hit. Their concussion rolled across the ridge, rattling loose snow from the rocks above them. The revetments went up in jagged showers of earth and splintered wood. Fires took hold, and the enemy positions that had seemed methodical and steady only minutes before convulsed and fell silent in places.

When the strike finished, Walker spoke into the radio. "Falcon Fire Control — consolidate. Commence fire for effect. Duration: thirty minutes. Keep patterns tight and focus on ammunition points and remaining guns."

Shellfire answered like thunder. The valley became a series of impacts: rounds feeding into one another, waves that smashed into dug-in positions, craters swallowing the roads and tracks that had fed the enemy logistics. LT White kept calling corrections and effect reports—secondary detonations, burning fuel, interrupted chokepoints. Davis fed updates to Walker, who coordinated adjustments and passed priority targets to the batteries.

Jack checked his watch as minutes bled into the steady volley. The smell of cordite and smoke cut through the crisp air. For thirty minutes the artillery pounded the valley, shaping the ground and the enemy's will.

When the barrage finally eased, three things were true: the gun pits were shattered, the supply tracks had been ripped to pieces, and the valley had been transformed from a staging area into a field of ruin. Walker let out a long breath and turned to Jack.

"You did good," he said simply. "You put eyes on the right place. We'll hold this and follow up. Get ready — we may need to move into the valley to exploit the damage."

Jack's jaw tightened in exhaustion and relief. He glanced back at his men—cold, tired, but alive. The price had been counted in explosions and risk, but for the line below, the storm they'd called down might have bought hours, maybe days.

LT White rubbed his hands together and checked his notes. "We'll monitor for any counter-moves. They'll try to shift, withdraw, or—if they're stubborn—re-mass elsewhere. Keep your heads down."

Chapter 18 Orders and Endings

The smoke still hadn't cleared.

From their perch on the ridge, Jack and the others watched the valley below writhe with fire and ruin. The snow that had once blanketed the ground was now blackened, churned into steaming mud. Craters overlapped like scars, each one still smoking from the thirty-minute barrage that had torn through the Chinese lines.

Even from this distance, they could see the damage. Gun pits were smashed open, barrels twisted, supply sleds overturned and burning. Bodies and broken equipment littered the ground in chaotic lines. The heavy silence that followed the storm of fire was almost worse than the barrage itself.

CPT Walker lowered his binoculars, his voice low. "Hell of a sight."

LT White, crouched beside him, adjusted his own field glasses and nodded. "Fire mission landed exactly where it needed to. Artillery pits were destroyed, ammo points burning out. But they'll regroup, sir. They always do."

Jack was silent for a long moment, scanning the wreckage through his scope. "They're moving," he said finally. "Back toward the tree line, east ridge. Small groups, maybe survivors or reserve teams. They're trying to pull back whatever's left."

Walker's jaw tightened. "And we're going to make sure they don't get the chance to rebuild." He turned to face Jack. "Mercer, I want confirmation of what's left down there—guns, equipment, bodies, comms wire, anything useful. Take your team down into that valley and bring me back a complete picture. If they've abandoned positions, I want to know why."

Jack lowered his binoculars and met Walker's eyes. The smoke made everything look like a different country—distorted, dangerous—and Jack didn't like sending the whole squad into that mess if he could avoid it.

"Sir," he said quietly, "I'm not taking the whole team down. Two men and I can move cleaner than six. I'll take Davis with me — he's my comms and steady as anything — and we'll pick our way through the wreck to confirm what's left. The others stay in the hide and hold the ridge. If anything goes wrong, they provide the exfil and cover."

Walker considered him for a long beat; pencil stalled over the map. "You sure you don't want Hill and Kim with you? More eyes, more hands."

Jack shook his head once. "Two bodies are easier to hide and quicker to move. Hill and Kim keep your approach secure and can hustle to our rally if we need them. I want the valley to be walked clean, undetected. Too many men make noise, and noise brings the wrong kind of attention."

White glanced between them, expression tight. "You'll need firm comm discipline. If you're that deep and close to wreckage, the enemy might be listening for movement."

"I know," Jack said. "Davis runs the net. We'll check, report, and get out. If we find anything actionable we'll signal. If we trigger something, the men on the ridge keep eyes on us and call suppression."

Walker rubbed his jaw, then nodded once. "All right. You and Davis go in. Hill and Kim hold here with the rest. I'll coordinate the overwatch and Standby fire. Move when you're ready — and Mercer, bring them back."

Jack dipped his head. "Yes, sir."

He handed the binoculars back to Walker, crouched, and moved to Davis. The two of them double-checked gear in the half-light while the others melted into the lee of the rock and the snow cave, preparing to wait and watch. Then, like ghosts, Jack and Davis slipped down into the smoke and the ruined valley below.

They moved like shadows through the shell-torn valley, boots sinking into wet, blackened snow where the bombardment had stripped away the white and churned the earth into slush. Smoke hung in the hollows in thick curtains, turning the world a flat gray. Every inhale brought the acrid bite of cordite—and underneath it, a new, fainter, fouler scent that made Jack's stomach tighten: the sick-sweet stench of something burned human, a smell that clung to the throat and wouldn't wash out.

The wreckage was worse than Jack had expected. Emplacements had been ripped open barrels twisted, revetments collapsed into ragged mouths, ammunition crates torched into smoking husks. Tracks that had once threaded neat lines into the enemy logistics belt were now ruts and gashes. Torn canvas and shattered sled runners littered the ground. Here and there, a dark, scorched shape lay half-buried in the snow, the outline of a man flattened against the earth; Jack looked away and kept moving.

They worked methodically. Jack moved from pit to pit, scavenging what could be useful— dog-eared ledgers half-charred but legible, a rain-sodden manifest wiped with his sleeve, a clipped radio handset wrapped in oilcloth. Davis photographed key positions on a small camera and scribbled grid references in quick shorthand.

Anything that might reveal intent—fuel markings, convoy times, unit sigils—went into Davis's notes.

At the flank a pair of figures slipped between burned tents and a collapsed revetment. Not a coordinated counterattack—more like patrols stumbling through the ruins, disoriented by the strike and the smoke. Jack flattened himself into a shallow crater and watched them pass within thirty yards, rifles slung but alert. They moved slowly, heads low, bewilderment on their shoulders. No shout. No sweeps. Just men trying to make sense of sudden ruin.

Davis's voice was barely audible in Jack's ear as he put a hand to the handset. "Mercer, I've got photos of the primary pits and the supply track. Sending grids now." He keyed the transmit switch and sent the coordinates with the practiced rhythm of someone who'd done this a half-dozen times. A crackled reply came—CPT Walker's voice sharp with static—acknowledged. Maintain eyes. Do not engage unless necessary.

They skirted a long line of overturned sleds and found the motor pool—if it could still be called that—smashed under a tangle of timbers. In a dugout beneath a collapsed tarpaulin the smell of fuel hung thick; nearby, the smell of burned flesh rose again in a hot, metallic wave that made Jack taste iron. He pried loose a damp manifest, wiping soot from the ink with his sleeve, and read aloud while Davis copied inventory into his book: fuel drums marked with unit sigils, crate counts— enough to move a reinforced company if it had survived. Now it was ash.

"Walker," Jack breathed into the throat mic when Davis had the net open. "SSG Mercer. Effects observed — gun pits destroyed, ammo points burned, supply nodes rendered inoperable. Sending grids for primary positions. No usable fuel caches remain at these coordinates. Over."

Static, then Walker: "Copy SSG Mercer. Good work. Hold and report any survivors; if you can, capture one for ID. We need confirmation of unit designation and intent. Proceed with caution — I'll keep the batteries on standby. Out."

Jack scanned the ruined lines and then looked at Davis. "We'll take one back. It'll be noisy, but it's worth the intel." He nodded toward a shallow slit beside a knocked-over revetment where a single enemy soldier was stumbling, coughing, clutching a damaged pack. The man's uniform was muddied and mismatched—clearly a straggler, dazed, probably not expecting prisoners.

They closed the distance using shell-holes for cover, moving slowly and low so their profiles cut against the broken ground. Jack slid up behind a charred crate while Davis swung wide to cut off escape. The prisoner didn't see them until Jack's hand clamped the back of his collar and the muzzle of the carbine rested at the base of his skull.

One clean smack to the back of the head—no theatrics—and the man folded. Davis was already there with waxed cord, binding hands tight and gagging him with a strip of cloth. The prisoner's eyes were wide and uncomprehending, breath shallow. Behind them, a pair of disoriented patrolmen passed within earshot and kept moving; perhaps they assumed the bound man was dead or simply overlooked him in the chaos.

Davis whispered into the mic: "Mercer to CPT Walker. Prisoner secured at grid three-four-five-eight-one-two. Requesting exfil route to friendly lines on the far side of the valley. We'll move with minimal noise and meet at the eastern perimeter. Over."

Walker's reply was crisp. "Copy. I'll coordinate extraction and notify CP. Move fast—keep to shell-holes and smoke for concealment. Trucks will rendezvous at the far-line point. Hold radio silence unless you get in contact. Out."

Jack slung the small bundle of documents and checked the prisoner—still breathing, hands bound. "Davis," he said, voice low, "you've got nets and the manifest. Start the timeline: trucks, convoy times, emplacement coordinates. We move on my mark."

They slipped through the wreckage, stepping in narrow paths between craters, using smoke and broken terrain as cover. Sporadic patrols passed again—two here, three there—men bewildered and moving to pull survivors and salvage what they could. Jack kept them unseen, letting the natural confusion do the concealment.

Halfway to the far-side wire, Davis hissed: "One more patrol, two o'clock, moving fast." Jack signaled them flat, and they went still. The patrol's boots crunched within ten meters; for a long minute the only sound was frozen leather creaking. The patrol paused to check a burned sled and moved on unnoticed. Jack's chest felt tight; each step was a bargain with luck.

When the friendly perimeter came into view—sandbags and the dull glow of lamps—they heaved and slipped through the gap where engineers had folded the wire.

They moved toward the command post, the valley behind them still glowing in places—embers flaring red against the snow. The stench clung to the air: cordite, burned oil, and the unmistakable, stomach-turning reek of charred flesh. It hung in their clothes, in their hair, in their lungs. Davis tried to keep his breathing shallow, but it didn't help. The smell was everywhere.

At the perimeter, two MPs stood waiting beside the sandbagged entrance, rifles slung and eyes wary. When they saw the bound prisoner between Jack and Davis, they straightened immediately.

Jack handed the man over without a word. "Hold him here," he ordered. "CPT Walker will have G-2 take custody. He's still disoriented—don't let him talk to anyone."

One of the MPs nodded and yanked the prisoner to his knees in the snow, checking his bindings. The man looked up once, dazed and hollow-eyed, before his head sagged forward.

Jack turned and stepped into the bunker with Davis close behind. The sudden warmth hit like a blow after the frozen air outside. Lanterns cast dim pools of yellow light across maps and radios. CPT Walker was hunched over the table, headset slung around his neck, talking quietly into the field phone. When he spotted them, he motioned them over and hung up.

Jack placed the small bundle of documents, notebooks, and captured maps on the table. "Recovered from the artillery positions, sir. Supply manifests, movement logs, and what looks like an order of battle for at least one regiment. All marked and dated within the last week.”

Walker flipped through the pages, his expression tightening with each sheet. “And the prisoner?”

“Outside, sir. MPs are holding him. He’s in rough shape but alive. G-2 should get something out of him once he’s warmed up.”

Walker nodded once. “Good. You did the right thing to keep him out there. No need to bring that smell inside.” His gaze lifted to Jack’s face, reading the exhaustion there. “You’ve seen enough for one day.”

Jack hesitated. “Sir, there’s nothing left in the valley. Everything—guns, ammo, fuel—it’s all gone. The artillery tore it apart. Whatever the Chinese were planning down there isn’t happening now.”

Walker’s jaw tightened as he set the papers down. “Then you’ve done more than your share. That strike may have just bought us a few weeks of breathing room.”

He leaned back, rubbing a hand over his eyes, then looked at both men. “You’ll file your debrief in writing after chow. For now, get cleaned up and get some sleep. We’ll start again when G-2 finishes with the prisoner.”

Jack and Davis nodded, saluted, and turned for the exit. The cold hit hard when they stepped back outside. The prisoner still knelt in the snow between the MPs, shivering, head bowed. For a moment, Jack's gaze lingered on him—one survivor in a field of ash—and then he turned away.

Behind them, CPT Walker's voice rose again inside the bunker, steady and sure, already passing their coordinates and findings up the chain. The valley was still burning, but the war was moving on.

The canvas tent was dim and heavy with the smell of oil and cold canvas when Jack and Davis pushed through the flap. Hill, Kim, and Reilly were already there, their gear laid out in tidy lines, weapons cleaned and reassembled. The lantern on the table cast a low, yellow light that caught the sheen of gunmetal and the steam from a half-empty coffee tin.

Hill looked up first. "Well, damn. Thought you two might've frozen solid out there."

Reilly grinned, though it didn't reach his eyes. "You look like hell, Sarge. You find the end of the world down there?"

Jack didn't answer right away. He unshouldered his pack, the weight of it landing with a dull thud. Davis followed suit, setting the radio down beside the cot before rubbing the cold from his hands.

Kim stepped forward. "You want a hand with your gear? We're already squared away."

Jack shook his head. "No. You did your part. I'll handle mine." He sat on the edge of his cot, pulled the M1A1A1 Paratrooper Carbine from its sling, and began to field strip it with slow, practiced motions. The faint clicks of the bolt and pins filled the tent. The weapon was filthy—ash and grime caked in the receiver, traces of burned powder in every groove. He worked it clean, methodically, his hands steady even when his thoughts weren't.

Hill exchanged a look with Kim, then finally asked, "What was it like down there?"

Jack kept his eyes on the rifle. "Bad. Worse than I've seen in a while. Nothing left standing. They were cooked in their own trenches." The tent went quiet after that, the only sound the rasp of his cleaning rod through the barrel.

Reilly poured two cups of coffee and handed one to Davis. "You boys earned it."

Davis nodded his thanks, eyes half-closed from exhaustion. "Walker's got G-2 on the prisoner already. Whatever that guy knows, they'll dig it out of him soon enough."

Jack didn't respond. He snapped the rifle back together, checked the action, then leaned it beside his cot. For a moment, he just sat there, elbows on his knees, staring at the floorboards. The room was quiet except for the soft ticking of the lantern and the faint wind outside.

The Misfits knew better than to press him further. They'd all seen that look before. So they went back to their routines—Hill rolling his shoulders under his blanket, Kim checking the laces on his boots, Reilly scribbling notes in his pocketbook—each man pretending not to hear the silence settle between them.

Jack finally spoke, voice low but steady. "Get some rest. Walker's not done with us yet."

No one argued. The lantern dimmed, the cold pressed back in, and for a few hours, the Misfits let exhaustion win. But none of them really slept—not with the memory of the valley still burning behind their eyes.

The command post was quieter than usual when Jack stepped inside. The lamps burned low, the hum of the radio steady and distant. CPT Walker stood at the map table, his sleeves rolled up and a steaming mug beside him. He looked up as Jack entered, gave a nod toward the empty chair across from him.

"Have a seat, Mercer."

Jack sat, resting his M1A1A1 Carbine beside him, the weapon still as much a part of him as the uniform. Walker studied him for a long moment before speaking.

"The prisoner you brought in," Walker began, his voice even, "turned out to be more than we expected. Company commander—frontline unit from the 116th Regiment, 39th Chinese Division. G-2 pulled a lot out of him once they got him talking."

Jack leaned forward slightly. "What did he give up, sir?"

Walker tapped the ash from his cigarette and pointed to a marked section on the map. "They're not preparing to launch another offensive. They're digging in—defensive positions in depth, layered strongpoints, fallback lines. The whole valley north of the thirty-eighth is being turned into a wall. Their orders are to hold us south of the Yalu and keep the Allies from advancing back into North Korea."

Jack exhaled slowly, absorbing the weight of that. "So, it's going to be a stalemate."

"Looks that way," Walker said. "At least for now." He paused, his tone softening. "Your team's work helped prove it. The intel you brought back from that valley—those coordinates—confirmed what the prisoner said. Higher knows you've been pulling more than your share."

He reached into his breast pocket and unfolded a single sheet of orders. "You and your team have been recommended for commendation. Division's been watching you for months now. The 2nd Infantry Division Commander himself signed off on something special." A faint grin tugged at the corner of his mouth. "Two weeks of R&R in Japan. You leave tomorrow morning."

For the first time in weeks, Jack let himself breathe a little easier. The thought hit him hard and warm—Patricia. He'd be able to see her again. Maybe even touch her hand without wondering if it would be the last time.

But Walker wasn't done. He straightened, his expression turning more serious. "There's more, Mercer. The Army's deactivating the Ranger companies. Command's reorganizing everything for the next phase of the war. Everyone's being reassigned—most to the 187th RCT, Airborne."

Jack felt the words sink in, cold and slow. "And my team?"

Walker's gaze softened. "You and your Misfits won't be going with them. Division G-2 wants you. They're standing up a recon section attached directly to 2nd ID Headquarters. You'll report to me until the transfer's finalized, then you'll work with their intelligence section. Eyes forward. Deep recon. You'll be the ones spotting the next move before anyone else knows it's coming."

Jack nodded slowly, the weight of change settled in. "When do we move out, sir?"

Walker smiled faintly. "Tomorrow. Transport leaves at 0700. You'll fly out of Pusan to Tokyo for your R&R before reporting in."

Jack rose, the faintest hint of relief creeping into his voice. "Understood, sir."

Walker extended his hand. "You've earned it, Mercer. Tell your men to get their gear ready and enjoy the time off. Something tells me the next chapter of this war won't give us much rest."

Jack shook his hand firmly. "Thank you, sir."

When he stepped back out into the cold night, the stars were sharp against the dark. For the first time in months, the horizon didn't look like another battlefield—it looked like a promise. Patricia's face flickered in his mind, her smile soft and steady, the warmth of her touch still as vivid as the day he'd left.

Tomorrow, he thought. Tomorrow, I'll see her again.

Chapter 19 Between Battles

The squad tent was quiet when Jack pushed through the flap. The lantern inside threw long shadows over the canvas walls, and the Misfits were where he'd left them—gear stowed, weapons clean, the smell of oil and coffee hanging in the cold air. Hill looked up first, reading Jack's face before he even spoke.

"Bad news, Sarge?" he asked.

Jack didn't answer right away. He stood there for a moment, letting the flap fall shut behind him. The warmth of the tent was faint, almost fragile after the chill of the command post. He hung his helmet on the cot post, unslung his M1A1A1 Carbine, and finally said, "Not bad. Just… different."

The men straightened, curiosity cutting through the fatigue.

Jack took a slow breath. "Walker just got word from G-2. The prisoner we brought in. He was a company commander with the 116th Regiment. He told them the Chinese are digging in—building defensive lines in depth across the northern valleys. They're setting up to hold us south of the Yalu, maybe for the long haul."

Kim frowned. "So, the fight's not over—it's just changing."

"Exactly," Jack said. "They're fortifying everything. That means less movement, more trench work, and probably a long winter."

Hill muttered, "Hell of a thing to freeze for."

Jack's mouth tightened slightly. "Yeah. But there's something else." He reached into his pocket and pulled out the folded orders Walker had handed him. "Command's giving us two weeks of R&R in Japan. Division Commander himself signed off. We leave tomorrow morning."

For a heartbeat, no one said anything. Then Reilly let out a low whistle. "You're kidding."

Davis laughed under his breath. "Japan? Real beds, real food, maybe even hot water?"

"Don't forget beer," Hill added.

Jack smiled faintly, the first time he had in days. "You earned it. Every one of you."

But before the relief could sink in, his expression shifted again calm but serious. "Walker also told me the Rangers are being deactivated. We're done as a unit."

The tent went silent. Even the wind outside seemed to pause.

Jack continued, his tone steady. "Most are being reassigned to the 187th RCT Airborne. But not us. G-2 requested the Misfits by name. We're being transferred to Division Headquarters. We'll be working recon directly for intel—deep ops, target observation, infiltration. Different kind of fight, but the same purpose."

Kim looked down at his boots, then nodded slowly. "Guess we're not done yet."

"Not by a long shot," Jack said quietly. "But for the next two weeks, we rest. We get our heads clear. We'll need that before whatever comes next."

Reilly leaned back on his cot, shaking his head. "Two weeks in Japan... I almost forgot what peace felt like."

Jack said nothing, but inside, the thought warmed him. Patricia. Her face, her voice, the quiet steadiness that had kept him grounded through it all. For the first time in months, there was a path leading back to her.

"Alright," he said, breaking the silence. "Get your gear squared away. We roll out at 0700. Next stop—Japan."

The tent filled with quiet motion—packs being checked, boots laced, and for once, low laughter in the cold air. Outside, the wind whispered across the camp, but for the Misfits, it finally sounded like something close to peace.

The engines growled low as the trucks idled near the airstrip, their exhausts curling into the frozen dawn. The Misfits climbed aboard the transport plane one by one, breath steaming in the cold, packs slung over their shoulders and rifles tight against their chests. The air smelled of diesel, oil, and winter. Jack was the last to climb the ramp, his boots thudding softly against the metal as he glanced once over his shoulder—the hills of Korea still shrouded in mist and smoke.

Inside, the C-47 was dim and loud, canvas seats stretched along the fuselage walls. Davis settled near the rear with the radio cradled in his lap, Hill and Kim trading low jokes that barely carried over the drone of the engines. Reilly leaned back, eyes already closing, the exhaustion of three months catching up at last.

Jack sat by the small oval window, the M1A1A1 Paratrooper Carbine resting across his knees. The engines roared, the ground shuddered beneath them, and as the aircraft lifted off, the world outside began to fall away—first the runway, then the camp, then the torn ridgelines and the scarred valley beyond.

For a moment, he allowed himself to breathe. Then the memories crept in.

The vibration of the engines became the rumble of another plane—one over occupied France. The cold on his face was the same, the smell of metal and oil identical. But instead of the Misfits, there were scientists and soldiers in gray uniforms, notebooks instead of rifles. He remembered the lights, the restraints, the sound of voices that spoke German too quickly for him to follow. The sting of the needle. The cold fire that crawled through his veins.

He saw flashes—white light, the sense of falling, the burn that wouldn't fade. He remembered the officer with the thin smile telling him it was all for "the advancement of the Reich." The smell of antiseptic and blood filled his lungs. The world had gone dark and then stretched, time bending until days blurred into hours, hours into years.

A sudden lurch of the aircraft snapped him back. Hill was laughing about something, Reilly groaning in mock protest. Jack blinked and forced the ghost of the lab back down where it belonged. He flexed his hands—steady, warm, alive—and looked out the window again.

The coastline slid by beneath them now, the sea stretching wide and gray toward Japan. Somewhere beyond that horizon was Patricia. She was the only constant in all the chaos, the only proof that he was still human beneath the scars and secrets.

He leaned his head back against the vibrating fuselage, eyes half-closed. The hum of the engines steadied his breathing. For now, the war was behind them. For now, there was peace—fragile, temporary, but real.

Tomorrow, he would see her again.

The engines wound down with a slow metallic sigh, the sound fading into the still morning air. The Misfits stepped out of the transport one by one, blinking against the soft sunlight breaking through the mist that rolled across the runway. Japan smelled different—cleaner somehow, salt air mixed with jet fuel and something faintly floral drifting in from the nearby trees.

Jack adjusted the strap of his pack and took a long breath, letting the tension of the flight drain from his shoulders. Davis came up beside him, shading his eyes as he looked across the tarmac. "Never thought I'd see this place again," he said quietly.

"Enjoy it," Jack replied. "We've earned it."

A sergeant with the R&R compound staff waved them toward a waiting jeep convoy. The drivers wore crisp uniforms and easy smiles—the kind that belonged to men who hadn't been living in frozen foxholes for months.

"Welcome to Camp Motoshima, gentlemen," the sergeant said as they approached. "You'll be billeted here for the next two weeks. Division sent word ahead—you're to get priority intake."

The Misfits piled into the jeeps. As they rolled through the base, the contrast was jarring. Where Korea had been mud, smoke, and the constant crack of distant artillery, Japan was alive with motion—supply trucks, laundry lines, the hum of radios playing faint music in the distance. The buildings were solid, the air warm. For the first time in months, it didn't feel like the world was trying to kill them.

The convoy stopped near a row of Quonset huts freshly painted in dull green. Inside, clerks in khaki guided them through the intake process— medical checks, forms, signatures. A young lieutenant briefed them on base regulations and recreation options, his voice sharp and practiced.

"You'll be assigned to Section B for your stay," he said, flipping through a clipboard. "Mess hall's open from 0600 to 2100. There's a club down by the shore—beer, cards, and if you're lucky, hot food that doesn't come from a can."

Hill gave a low whistle. "Hot food and beer? I might just stay."

The lieutenant ignored the comment and moved on. "You'll also turn in your weapons and combat gear for storage. No arms inside the compound."

That drew a few grumbles, but no one argued. The Misfits lined up by the armory tent,

each man stepping forward to surrender his weapon.
The armorer checked serial numbers and tags before
stowing them in a rack marked "2ID G-2 –
Detached."

Jack hesitated when it was his turn. He
looked down at the M1A1A1 Paratrooper Carbine
resting across his palms—the same weapon that had
been with him since Normandy. The walnut grip
was worn smoothly where his fingers always rested.
It wasn't just a rifle; it was a part of him.

The armorer cleared his throat. "Sir?"

Jack gave a small nod, then handed it over.
The armorer tagged it carefully, placed it in the
rack, and saluted. "She'll be here waiting for you
when you're back on duty."

Jack didn't answer, just returned the salute
and stepped away.

When the last weapon was turned in, the
Misfits were led to their quarters—a long, narrow
hut with bunks lined along the walls and clean
sheets that smelled faintly of soap. Davis dropped
his pack with a sigh. "Hell, I forgot what a bed even
felt like."

Kim sat on the edge of his bunk, staring out
the window toward the harbor where ships gleamed
in the sunlight. "Feels wrong," he said softly. "Like
we left part of the war behind us."

Jack stood in the doorway for a moment, watching his men settle in. They were alive, whole, and for the first time in months, safe. It was strange—almost unsettling—to feel that kind of quiet.

He walked to his bunk, sat down, and let his hands rest on his knees. For the first time in a long while, he didn't feel the weight of his rifle. Just the weight of everything that had brought them here.

Two weeks of peace. Two weeks before the next storm.

The barracks were quiet after intake, the hum of the compound muffled by the sound of the sea outside. Jack stood by the open window for a long moment, the scent of salt and diesel on the breeze, then finally turned back to face his men.

"Alright, listen up," he said, his voice steady but lighter than it had been in weeks. "We're here to rest, but that doesn't mean we go soft. Hit the showers, get cleaned up, and enjoy the day. The war's still out there, but for now, we've got hot water and real food. Make the most of it."

Hill raised an eyebrow. "You mean… actually sleep in? Not dig a foxhole?"

Jack almost smiled. "Yeah, that's exactly what I mean. But starting tomorrow morning, at 0600 sharp, we'll do PT. Nothing crazy—just keep sharp. Run, stretch, stay in shape. After that, the rest of the day's yours. Eat, sleep, play cards, whatever you need."

Reilly let out a low chuckle. "You heard the man—back to being human again."

Kim and Davis were already pulling clean uniforms from their duffels, the rare sight of relaxed grins crossing their faces. For once, there was no rush, no alarms, no orders snapping in their ears.

Jack grabbed a towel from his pack and slung it over his shoulder. "I'll see you all in the mess hall later. Try not to get into trouble."

Hill smirked. "No promises."

Jack shook his head and stepped out into the warm light. The compound was quiet except for the distant sound of waves and laughter somewhere near the beach. It felt wrong to hear laughter again.

He walked toward the shower block, boots crunching softly on the gravel. Steam drifted out from the open windows, carrying the scent of soap and heat. As he stepped inside and the first rush of hot water hit his skin, months of grime, blood, and frost began to wash away. He closed his eyes and exhaled slowly.

For the first time since landing in Korea, he allowed himself to think about something other than survival—Patricia. She was here, somewhere on this base, working at the hospital. Just a few buildings away, maybe. The thought steadied him.

Tomorrow, he told himself. After the team's settled. Tomorrow, he'd see her again.

The mess hall was alive that evening, laughter echoing under the low tin roof as the Misfits crowded around a table near the back. Real food—eggs, bread, and something that vaguely resembled steak—filled their trays. Reilly was already halfway through his second helping, grease shining on his chin.

"God bless whoever cooked this," he muttered. "I don't even care what it is."

Hill grinned. "That's because it's not powdered or canned. Feels like a holiday."

Kim leaned back in his chair, a rare, easy smile crossing his face. "First night in months we're not freezing or being shot at. I'll take it."

Jack sat quietly with his coffee, watching them relax. The exhaustion, the constant tightness in their shoulders—it was finally starting to fade. He didn't join the jokes much, but every time one of his men laughed, he felt something loosen inside. For the first time since Korea, they looked human again.

After chow, they lingered near the rec hut—cards, cigarettes, stories that grew taller with every retelling. The night air was mild, a soft sea breeze rolling in from the bay. Jack stayed a while, smiling faintly as Davis and Reilly argued over a poker hand, before finally slipping away toward the barracks. Tomorrow would come early, and he wanted his team ready for it.

The sun was just beginning to climb when the Misfits hit the track running. Their breath came in steady clouds as boots pounded the packed dirt, the rhythm of motion breaking through the morning quiet. Davis set the pace, Hill grumbling behind him, Kim silent and steady as always.

Jack ran a few paces ahead, his mind half on the cadence, half on the hospital he knew was just beyond the next row of barracks. He hadn't slept much—too busy replaying what he'd say when he saw her again.

As they rounded the corner, the hospital came into view—whitewashed walls gleaming in the sunlight, a line of nurses moving between buildings. And then he saw her.

Patricia.

She stood near the entrance, clipboard in hand, sunlight catching the faint copper in her hair. For a moment, Jack thought he was imagining it. Then she turned—and froze. Her eyes widened, and the clipboard nearly slipped from her fingers.

"Jack?" she breathed. Then louder, disbelieving: "Jack Mercer!"

He slowed to a stop as the rest of the team thundered past, giving each other knowing looks as if they disappeared down the path. Jack managed a crooked grin. "Morning, ma'am."

She crossed the space between them in seconds; her voice firing questions almost faster than he could answer. "What are you doing here? When did you get back? How long have you been on base? Why didn't you come find me sooner?"

Jack raised his hands slightly, laughing under his breath. "I was going to—swear to it. We just landed yesterday. I figured I'd shower first before showing up in front of you smelling like the DMZ."

Her expression softened, though the corners of her mouth trembled with emotion. "You idiot," she whispered. "You should've come last night."

"I wanted to," he said quietly. "But I needed to make sure the team was settled. We've got some breathing room now."

Patricia stepped back just enough to look at him properly, eyes shining. "When can I see you?"

Jack hesitated for only a moment. "When do you get off duty?"

"Eleven hundred," she said without thinking. "Meet me in the garden—where we met before you shipped out."

Jack's smile deepened, the memory flashing bright and warm. "Twelve hundred it is. I'll be there."

She nodded once, a small, shaky laugh escaping her. "You better be."

As she turned back toward the hospital, Jack watched her go, heart hammering harder than it had in any firefight. The world around him—the base, the sun, even the distant surf—felt sharper somehow, more alive.

He took a slow breath, then started jogging to catch up with his men, a faint grin tugging at his lips. For the first time in years, he wasn't running toward a battle. He was running toward something good.

The garden behind the hospital was just as Jack remembered it—quiet, green, and tucked away from the noise of the base. A stone path wound between trimmed hedges and small lanterns, the scent of jasmine and salt air mixing on the breeze. It was peaceful in a way that felt foreign after so many months of war.

Jack stood by the bench beneath the same cherry tree where he'd said goodbye to Patricia before shipping out. The memory played vividly in his mind: her hand in his, the unspoken fear in her eyes, and the promise he'd made to come back.

He turned as she appeared at the end of the path, sunlight catching in her hair. She was still in her nurse's whites, cap tucked under one arm, and for a moment, the world narrowed down to just her.

"Jack," she said softly as she reached him.
"You look... different."

He gave a small smile. "Guess the war will
do that."

They sat, the silence between them thick but
not uncomfortable. After a moment, Jack's jaw
tightened slightly, and he turned to face her.
"Patricia, before this goes any further… we're
going against regulations, you know. You being an
officer and me an enlisted man—it's not exactly
allowed."

Her brow furrowed. "An officer?"

He nodded, almost sheepishly. "You're a
nurse working in an Army hospital. That makes you
an Army officer, doesn't it?"

Patricia blinked, and then—despite
herself—laughed, a soft, disbelieving sound. "Oh,
Jack… no."

He frowned. "No?"

She shook her head, smiling now. "I'm not
in the Army, Jack. I'm a civilian nurse assigned to
an Army hospital through the Red Cross program. I
never enlisted, never took a commission. I just…
never thought to explain that to you before you
shipped out."

For a long moment, Jack just stared at her,
the confusion fading into relief. "So… you're
telling me we're not breaking any rules?"

Patricia's smile warmed, eyes softening. "No rules at all."

Jack leaned back on the bench, exhaling a low breath that came out half a laugh. "That's the best news I've heard in a long time."

She watched him for a moment, the tension leaving his shoulders, and reached over to take his hand. "You've carried enough weight for one lifetime, Jack Mercer. You don't need to carry regulations that don't exist."

He turned his hand to lace his fingers through hers, the rough calluses of his palm brushing against her soft skin. "Guess I just needed to hear it from you."

Patricia smiled faintly, the sound of waves and wind filling the quiet between them. "Then consider it official."

Jack chuckled under his breath. "That's one order I'll follow."

They sat in silence for a while, the garden around them whispering with life. For the first time in months, Jack felt something close to peace—something he hadn't realized he'd been chasing since the day he left her behind.

The path wound gently through the garden, the fading afternoon light spilling gold across the leaves. Jack and Patricia walked slowly, their hands brushing now and then but never quite holding, as though both were afraid to break the quiet. The air smelled faintly of sea salt and blooming jasmine, and the distant hum of the base barely touched this part of the compound.

Patricia glanced up at him, studying his face. The hard lines around his eyes, the tension in his jaw—it was all there, written in the way he carried himself. "You've changed," she said softly. "More than just a little."

Jack gave a slow, tired smile. "That's what war does. Korea's… different. Feels colder somehow, even when it's not."

They passed beneath the cherry trees, the sound of gravel crunching under their boots. Patricia let the silence stretch before she spoke again. "You saw a lot, didn't you?"

He didn't answer right away. His eyes were fixed on the horizon, where the ocean met the sky. "Too much," he finally said. "We did things out there most people wouldn't believe. Walked into places that didn't exist on a map, watched good men disappear in the snow. You don't really come back from that."

Patricia slowed, turning toward him. "And yet here you are."

Jack's mouth tightened. "Physically, sure. But sometimes it feels like I left part of myself behind each time—Normandy, the Ardennes, now

Korea. It's like I'm still scattered across all of them."

She reached out and took his hand this time, firm and deliberate. "Then maybe you need someone to help gather those pieces."

He looked down at their joined hands, the roughness of his palm against her soft fingers. "You already are," he said quietly.

They walked on together in silence after that, crossing out of the garden and onto the narrow street that led toward the small set of apartments reserved for medical staff. The buildings were neat and whitewashed, their paper windows glowing softly in the evening light. Patricia's quarters were near the end, a small second-story apartment with a view of the bay.

When they reached the steps, she turned to face him. "Would you like to come in? Just for tea," she added quickly, though the warmth in her eyes said it didn't need to be justified.

Jack hesitated only a moment before nodding. "I'd like that."

As they climbed the steps, Patricia unlocked the door and held it open. Inside, the air smelled faintly of soap, lavender, and clean linen—a far cry from the canvas stench of army tents. A small kettle sat on the stove, a book half-read on the table. It was a space of peace, of normalcy.

Jack stepped in and paused, his gaze sweeping the room. "It's… nice," he said softly, the words carrying more weight than they should have.

Patricia smiled faintly as she set the kettle to boil. "It's home, for now. And I'm glad you're here."

Jack leaned against the wall, eyes softening as he watched her move about in the small kitchen. For the first time in a long while, he wasn't thinking about the next mission, or the next fight. Just the quiet sound of water beginning to boil, and the woman who'd waited for him through it all.

The room was quiet, the low hum of the kettle fading into the soft patter of wind against the window. Patricia set her teacup aside and crossed the small space between them. Jack hadn't moved— he sat where he'd been, eyes distant, the lines of war still etched deep into his face.

Without a word, she knelt beside him again, her hand brushing lightly against his cheek. "You don't have to keep talking," she whispered. "You've said enough."

Jack caught her hand and held it there, his rough fingers trembling just slightly against her smooth skin. "It's just hard to stop sometimes," he said. "It's all still so close."

Patricia stood, her gaze never leaving him, then sat gently on his lap, her arms sliding around his neck. The motion was natural—quiet, steady, like something inevitable. Jack's breath hitched, surprise giving way to something softer as he met her eyes.

"Then let it be far for a while," she said.

The words broke whatever barrier was left between them. Jack reached up, his hand resting against her back, feeling her warmth through the thin fabric of her blouse. She leaned in, their foreheads touching, the scent of lavender and soap filling the small room.

When their lips met, it wasn't desperate—it was slow, searching, the kind of kiss that carried months of distance and longing. The world outside faded until there was only her heartbeat against his chest and the gentle rhythm of their breathing.

Patricia's fingers traced the lines of his face, the scars, the stubble, the quiet strength. Jack's hand came up to cradle the back of her head, his thumb brushing through her hair. They stayed like that for a long moment, no rush, no urgency, just two people finding their way back to something human.

When they finally pulled apart, she rested her head on his shoulder, her voice a whisper. "You don't have to be the soldier right now, Jack. Just be you."

His arms tightened around her, a small sigh escaping him. "That's easier when you're here."

Outside, the sun dipped lower, painting the room in gold. Inside, the war and the world seemed very far away.

The light outside had dimmed into twilight by the time Jack glanced at the clock on Patricia's wall. He exhaled quietly and shifted, brushing a strand of her hair from her face. "I should get back to my men," he said softly. "They'll start wondering where I wandered off to."

Patricia smiled faintly, though there was a trace of reluctance in her eyes. "Duty calls."

He nodded. "Always does. But I'll still be here for another thirteen days. Two weeks of peace, Walker said."

Her smile grew a little warmer at that. "Then let's make those days count."

"How do you mean?" he asked, standing as she followed him to the door.

"I'm off duty at eleven tomorrow," she said. "If you're free, maybe we could go see some of the city—the markets, the harbor. There's a temple on the hill with a view of the whole bay. It's beautiful this time of year."

Jack hesitated only a moment before nodding. "I'd like that. A lot."

Patricia stepped closer, her hand resting lightly against his chest. "Then it's a date," she said, her voice low and certain.

He smiled, leaning in to press a soft kiss on her forehead. "Tomorrow at twelve," he said. "Don't be late."

She laughed quietly. "I won't be. Go on—your men need their sergeant."

Jack turned toward the door, pausing once to glance back. Patricia stood framed in the warm lamplight, watching him with the same quiet strength she always had. He gave her a small nod, then stepped out into the cool evening.

The air smelled of sea and rain, and for the first time in months, Jack felt something that almost resembled peace. Thirteen days in Japan. Thirteen days of something like normal. He didn't know what the future would hold—but for now, that was enough.

Chapter 20 The Furnace

The days that followed fell into a rhythm unlike anything Jack Mercer had known since before the war. Each dawn began the same way—boots pounding against the gravel path behind the compound, breath fogging in the crisp Japanese air as the Misfits ran in tight formation.

Jack led from the front, pace steady, voice calm and measured as he called cadence just loud enough to carry over the sound of the surf. The routine anchored them—pushups in the sand, stretches in the rising light, laughter that came easier now. No one complained. After months of cold, hunger, and fear, the sting of effort under a clear sky felt almost like freedom.

By midmorning, the team broke off to rest, eat, or wander into town. Jack gave them that space—they'd earned it. Davis spent hours tinkering with the radio gear in the rec hut, Reilly wrote long letters home, Hill and Kim sparred for fun behind the barracks. For once, there was no mission waiting to claim them, no sudden orders, no death marching beside them.

And for Jack, the rest of each day belonged to Patricia.

They walked the narrow streets of the coastal town, exploring markets that smelled of grilled fish and sweet bread. Patricia would stop to admire a kimono or a handmade trinket, and Jack would watch her with quiet wonder, still half disbelieving she was real, here, within reach. They climbed the temple hill one afternoon, the wind sweeping her hair loose as the bay glittered below them.

Conversations came easily now—about her patients, about his men, about the world they both wished could stay this peaceful. Sometimes they sat together in silence, her head resting against his shoulder, the simple closeness saying more than words ever could.

Each evening ended the same: Jack walking her back to her apartment, her hand brushing his as they said goodnight under the glow of the lanterns. And each night, as he made his way back through the quiet base, he carried the memory of her laugh, her touch, and the feeling that for the first time in years, the war inside him had gone still.

The eleventh day of R&R dawned calm and bright, the sea air drifting through the open windows of the barracks. Jack had spent the morning in silence, sitting on the edge of his bunk, watching sunlight glint off the water in the distance. His carbine was still locked in the armory, his boots polished, his uniform pressed—but for once, he wasn't thinking about the Army.

He was thinking about Patricia.

When he found her, she was standing outside the hospital gardens, a clipboard tucked under one arm, the faint wind tugging at her hair. She smiled when she saw him, the kind of smile that hit him harder than any firefight ever could.

"Jack," she said softly. "You're early."

He smiled faintly. "Guess I didn't want to waste time."

They walked together down the familiar path toward the little bench beneath the cherry tree. For a while, neither spoke. Then Jack stopped, turning to face her fully. The set of his jaw was firm, but his voice was quiet.

"Patricia," he began. "I've been thinking about this since the day I came back. About what comes next."

She tilted her head slightly, sensing the weight in his words. "What do you mean?"

He took a slow breath. "You know what I am—a soldier. I've been lucky to make it through this long, but luck doesn't last forever. And I can't stand the thought of something happening to me and leaving you without anything… without knowing you were taken care of."

Patricia's expression softened, though her eyes grew glassy. "Jack, don't—please don't talk like that."

He reached out, taking her hand. "I have to. I've seen what happens when men don't plan for what comes after. I don't want you to be another letter, another 'we regret to inform you.'"

Her breath trembled. "You think I haven't already thought of that? Every time I hear a report from Korea, every time a transport comes in with the wounded… I think of you."

Jack's thumb brushed across her hand. "Then marry me," he said simply. "Now. Before I go back. So that no matter what happens, you'll be taken care of. You'll have my name. My promise."

For a long moment, Patricia said nothing. The wind stirred the petals of the cherry blossoms, scattering them across the path. When she finally looked up, there were tears in her eyes.

"I don't want to think about losing you," she whispered.

He smiled faintly. "Then don't. Think about keeping me instead."

Patricia let out a shaky breath and nodded slowly, her fingers tightening around his. "Alright, Jack Mercer. Yes. I'll marry you."

Jack's heart felt like it might break from relief. He pulled her into his arms, holding her close as she buried her face against his chest. For a long while, neither spoke. The world was just the sound of the sea and the soft rustle of her breath.

By that afternoon, the Misfits were gathered outside the mess hall, wide-eyed and grinning as Jack broke the news.

"You're serious?" Hill asked, his voice breaking into a laugh. "You're actually tying the knot?"

"Tomorrow," Jack said with a calm smile. "You're all invited."

Reilly slapped Davis on the shoulder. "Guess that means we've got a wedding to get ready for, boys."

Kim raised an eyebrow. "You think the chaplain knows what he's in for?"

Jack chuckled. "He'll figure it out."

The next day, the small chapel on base was filled with soft morning light. The Misfits stood in the back, cleaned up and proud, as Patricia walked down the short aisle in a simple white dress borrowed from one of the nurses. Jack stood at the front in his freshly pressed uniform, his hands clasped behind him, heart hammering like he was stepping out of a plane again.

When she reached him, she smiled—a quiet, steady smile that said everything she didn't need to.

They spoke their vows softly, without flourish or spectacle. When the chaplain pronounced them husband and wife, Jack leaned in and kissed her, gentle and certain. Outside, the wind carried the sound of laughter and applause as the Misfits cheered and clapped, grinning from ear to ear.

Afterward, as the men offered handshakes and jokes, Jack turned to them with a rare grin. "Alright, Misfits—after today, you're on your own for the final stretch. I'm spending what's left of our R&R with my wife."

Reilly whistled. "Can't argue with that, Sarge."

Kim smirked. "Just don't forget we head back in two days."

"I won't," Jack said, his arm tightening around Patricia's waist. "I'll be there in the morning to draw my gear. But until then, I'm not thinking about anything but her."

Patricia leaned into him, her smile warm and sure. For the first time since the war began, Jack felt whole.

Two days later, the transport trucks rumbled across the compound yard, engines growling in the morning mist. Jack stood beside the Misfits as they went through equipment draw, the clatter of rifles, canteens, and web gear echoing under the tin roof of the supply shed.

He signed the last of his paperwork, the pen scratching against the clipboard, when Davis wandered over with a grin. "Good to have you back, Sarge. You miss all the excitement."

Jack arched an eyebrow. "Excitement?"

Hill groaned immediately. "Don't start."

Reilly was already laughing. "Oh, we're starting. You should've seen him, Jack. Hill here found himself a little romance while you were off playing newlywed."

Jack smirked, looking up from his checklist. "Is that right?"

Hill shot a glare at Reilly. "It wasn't like that."

Davis chimed in before Hill could defend himself further. "Was too! With a geisha girl from the teahouse by the harbor. He's been walking around like he's in a daydream ever since."

Kim grinned, tightening the strap on his pack. "Guess love's contagious around here."

The supply sergeant behind the counter grunted. "Long as it doesn't mess with your aim, boys."

The group broke into laughter, the sound easy and familiar. Even Hill couldn't keep his scowl for long. "Yeah, yeah," he muttered. "Laugh it up."

Jack shook his head, a faint grin tugging at his lips. "Well, Hill, at least you picked better scenery than I did last time I fell in love. Mine started with mud and artillery."

That got another round of laughter from the men. The tension of their return was already breaking, replaced by the simple comfort of banter and routine.

As they finished strapping on their gear, Jack looked over his squad—the same familiar faces, the same trust in their eyes. They were rested, fed, and ready.

He slung his freshly issued M1A1A1 Paratrooper Carbine over his shoulder and straightened. "Alright, Misfits. Make sure your gears tight and your packs are balanced. Trucks roll out in twenty. Korea's waiting."

Reilly groaned good-naturedly. "I was just starting to like this place."

Davis laughed. "You'll live. Maybe."

Jack smiled faintly as he checked his watch. He'd said goodbye to Patricia that morning, her kiss still fresh in his mind, her eyes the last thing he saw before the transport doors closed.

Now it was back to what he did best leading his men into whatever waited beyond the next horizon.

The morning air over the airstrip carried the sharp bite of jet fuel and sea saLT Rows of olive-drab C-47s sat waiting on the tarmac, propellers turning lazily as ground crews shouted over the roar. The Misfits stood in formation near the flight line, duffels slung over their shoulders, the hum of engines vibrating through their boots.

Reilly squinted up at the transport. "Feels like we just got here."

Davis grinned. "You say that every time we leave somewhere that serves real coffee."

"Yeah," Reilly shot back, "but this time I mean it."

Jack adjusted the sling on his M1A1A1 Paratrooper Carbine. "Enjoy the flight while you can, boys. When we hit the peninsula, it's back to work."

Hill sighed. "You mean back to waiting."

Kim smirked. "The war's standing still, but Walker's not gonna let us sit on our hands."

The crew chief waved from the base of the steps. "Mercer's team—Misfits, right? Load up on Bird Four-Two! Let's move—wheels up in ten!"

The men fell in line, climbing the metal steps one by one into the fuselage. The smell of oil and canvas filled the narrow space, the cold air vibrating with the hum of the twin engines. Jack took a seat near the front, his helmet on his knee.

When the C-47 lifted off, Japan shrank beneath them—green fields, tiled rooftops, and the gleam of the bay fading into mist. Jack watched through the small window until the coast vanished. His thoughts drifted to Patricia—her goodbye that morning, the warmth of her hand lingering against his. He tucked that memory away like a lucky charm and turned his mind back to duty.

Hours later, Korea appeared through low gray clouds, cold and unmoving. The war had reached a stalemate—neither side pushing, both entrenched and waiting. The C-47's wheels hit the frozen tarmac with a jolt, and as soon as the engines wound down, the cold rushed in.

The Misfits climbed down the steps into the biting wind. Trucks idled nearby, their exhaust hanging thick in the air. The men loaded up wordlessly, the laughter from Japan replaced by quiet focus as the convoy rolled north toward Division HQ.

By the time they reached the 2nd Infantry Division compound, dusk was creeping over the hills. The base looked unchanged rows of tents, sandbagged bunkers, the faint rumble of artillery miles away.

CPT Walker was waiting outside the command tent, clipboard under one arm. His uniform was dust-stained, his expressions steady and sharp as always.

"Mercer," he called as Jack jumped down from the truck. "Good to see you and your men back in one piece."

Jack saluted smartly. "Good to be back, sir."

Walker returned the salute with a faint grin. "Well, you didn't miss much. The war's ground to a standstill—neither side's moving more than a few yards in any direction. But that doesn't mean our work's done."

Hill muttered, "Figured as much."

Walker caught it and smirked. "You figured right, Hill. Command reorganized some of the staff while you were gone. I've been reassigned to Division G-2 to oversee all recon assets."

Jack raised an eyebrow. "So, you're running the recon teams now?"

Walker nodded. "That's right—and that means you still report directly to me. The Misfits are now part of 2ID Recon under G-2 operations. You'll be my forward eyes from here on out."

A murmur of satisfaction rippled through the team—different title, same commander, same trust.

Walker continued, "Get your equipment checked and settle in. I'll have a briefing for you tomorrow morning. The lines might be frozen, but intelligence never stops. We need to know exactly what the Chinese are digging in for."

Jack nodded firmly. "Understood, CPT Walker."

Walker gave him a steady look. "It's good to have you back, Mercer. Both you and your team."

As Walker turned and disappeared into the command tent, Jack looked over his men. The lighthearted air of Japan was gone—replaced once again by the calm, focused readiness that defined them.

"Alright, Misfits," Jack said, tightening the strap on his carbine. "You heard him. Let's get our gear squared away. The war may be standing still, but we sure as hell aren't."

The briefing tent was dimly lit, the heavy canvas walls rippling with every cold gust sweeping off the hills. A single lantern hung from a center pole, throwing long shadows over a map-covered table.

Jack and the Misfits sat on folding chairs, their faces serious and alert. CPT Walker stood near the front, flipping through a folder as he spoke.

"Alright, gentlemen," Walker began, "Division's looking at restructuring how we handle reconnaissance. We're—"

The tent flap opened, and the words died on his tongue.

Everyone in the room snapped to their feet, boots clapping together.

Major General Sims stepped through the entrance, the cold air following him in. He was a tall man with a lined face; his greatcoat dusted with frost. Every man straightened instinctively.

"At ease, men," Sims said, his voice deep and even. "Take your seats."

The Misfits obeyed, though none truly relaxed. A visit from the Division Commander wasn't routine.

CPT Walker straightened beside the table. "Gentlemen, this is Major General Sims, commanding officer of the 2nd Infantry Division. The General wanted to speak with you personally."

Sims nodded once, his sharp gaze settling on Jack. "Sergeant Mercer, isn't it?"

Jack stood immediately. "Yes, sir."

The General studied him for a moment, then gave a faint nod. "I've reviewed the reports from your missions over the past few months—especially the reconnaissance operations under CPT Walker's direction. You and your men have set a new standard for small-unit effectiveness. Quiet. Efficient. Precise."

Jack said simply, "Thank you, sir. We just do our job."

Sims allowed a small smile. "You do it better than most. The work you've done for this division has saved lives—more than you'll ever hear about. I want that kind of discipline and initiative multiplied across my recon elements."

He looked to Walker, then back to Jack. "That's why I'm here. We're forming two new reconnaissance teams modeled after yours. CPT Walker will oversee the entire program under G-2, and I want you and your Misfits to help him handpick and train the new squads. You'll screen the volunteers, test them, and build them up to your standard."

Jack gave a short nod. "Understood, sir. We'll find the right men."

"Good," Sims said. "We need teams that can move unseen, think for themselves, and bring back results. You've proven that works."

He retrieved his gloves from the table and met Jack's eyes once more. "Keep it up, Sergeant. The division's watching—and depending on you."

"Yes, sir."

Sims turned to Walker. "Captain, I'll expect your progress report within a week. Carry on."

As the general stepped out into the cold, the air in the tent seemed to ease. The men shifted, shoulders relaxing.

Walker gave a faint grin. "Well, Mercer, looks like you and your boys just got promoted to instructors."

Reilly leaned toward Davis and muttered, "Think that comes with extra pay?"

Walker's head snapped around. "No. But it comes with extra responsibility."

The Misfits chuckled quietly, and Walker turned back to Jack. "You heard the General. We start tomorrow. I want every volunteer screened—physically, mentally, and under stress. If they can't keep up with your team, they're out."

Jack nodded, the faintest smirk tugging at his lip. "Understood, CPT Walker. We'll find the best."

Walker's voice softened just a touch. "I know you will."

The command tent was nearly empty when the rest of the staff had gone. The lanterns burned low, casting amber light across the maps spread over the table. Outside, the wind howled faintly against the canvas, carrying the distant echo of artillery.

CPT Walker leaned over the map, coffee steaming in his hand. Jack stood opposite him, hands resting on the edge of the table, eyes tracing the markings that represented front-line positions.

Walker broke the silence first. "We've got a month, Mercer."

Jack looked up. "Sir?"

"One week to fill the other two teams," Walker said, his tone clipped and businesslike. "And three weeks after that to train them. Division wants all three recon elements ready to operate before the next round of offensive planning. That gives us thirty days, give or take."

Jack gave a short nod. "That's tight."

"Yeah," Walker said dryly. "So, we do it right the first time." He sipped his coffee, then set the mug down with a dull thud. "We need the right kind of men for this. Not just soldiers—recon men. Quiet thinkers. The type who doesn't freeze when they're alone in the dark."

Jack nodded slowly, considering. "We'll need a mix. Scouts, radio operators, maybe a medic for each team. And I want men who can move—no dead weight. Speed and stealth over brute strength."

"Agreed," Walker said. "Discipline, too. I don't want cowboys out there. We're not running commando raids anymore—we're running intelligence operations. The enemy lines are too tight for mistakes."

Jack leaned over the map, pointing at several zones marked in red grease pencil. "Terrain's going to make the difference. These hills aren't forgiving.

Whoever we pick must move silently and know how to read ground at a glance."

Walker nodded, eyes narrowing thoughtfully. "Alright. Tomorrow, we'll start the interviews. I want a list of volunteers from the line companies—infantrymen who've already been in the thick of it. No green replacements."

Jack rubbed his jaw. "We can test them in three stages. First, endurance—marching under load, weapons maintenance, obstacle course, timed. Then, fieldcraft—camouflage, movement, and observation. Finally, decision drills. See who keeps calm under stress."

A small grin tugged at the corner of Walker's mouth. "You've done this before."

Jack shrugged. "Learned the hard way, sir."

Walker chuckled quietly. "Fine. You run the course. I'll oversee and take notes. We'll pull the top ten and whittle them down to the six best for each new team."

"Roger that," Jack said. "And for the ones who make it—we'll train them the way the Misfits learned. Night ops, land navigation, radio discipline, and extraction drills."

Walker nodded approvingly. "Good. I want them operationally ready in twenty-one days. That means we push them hard from day one. No breaks."

Jack gave a short smile. "They'll curse us for it."

"They'll thank us when it keeps them alive," Walker replied, his voice quiet but firm.

For a moment, both men stood in silence, the weight of responsibility hanging heavy between them. Outside, the wind picked up, rattling the tent poles like distant gunfire.

Walker finally spoke again. "You've built something solid here, Mercer. The Misfits—they're not just good soldiers. They're a standard. These new teams will carry your mark. Make sure they live up to it."

Jack met his eyes. "They will, sir."

Walker reached for his coat, pulling it on. "Good. We start at 0700 tomorrow. You get some sleep—you're going to need it."

Jack gave a small nod as Walker stepped into the cold night. The tent flap fell closed, leaving Jack alone with the maps and the faint hiss of the lantern.

He looked down at the grid lines, the red markings, the drawn front. Then he exhaled slowly and muttered under his breath, "Time to build ghosts."

Dawn came cold and hard, the kind of gray that makes every breath visible and every movement look sharper. The training ground lay a few klicks from the division perimeter — a ragged stretch of ground where wind had stripped the grass down to the dirt, dotted with low ridges and boulder fields. Jack and the Misfits were already there before first light, setting markers, laying courses, and running final checks on the timers and compasses.

CPT Walker stood with a clipboard, the steam from his coffee steaming in the air. "We start at zero seven hundred," he said. "Two teams of six each. Endurance, fieldcraft, and decision drills. Keep it tight. No favorites. If they can't do the basics, they won't live out there."

Volunteers trickled in—men from various line companies, faces weathered by mud and cold, eyes trying to be confident even when their shoulders sagged from months in the line. Jack watched them as they assembled: the walk of some who'd been on foot for weeks, the nervous grin of a private who'd never been tested like this. Davis organized the radio bench; Kim and Hill checked harnesses and weighed rucks. Reilly set up the obstacle course with a crooked grin and a clipboard that said "Mercer Trials" across the top.

"Alright," Jack called, voice carrying. "Warm up. First test is a timed six-kilometer march with full kit. Two-hundred-meter sprint at the end. If you can't make it in forty minutes with a sixty-pound ruck, you don't come near my observation post."

The whistle blew and the line moved. Men hammered down the trail, boots kicking up clouds of dust and gravel. Jack jogged the course with them for a stretch, watching gait and breathing, how they adjusted straps, how they re-set a limp shoulder without stopping the pace. One man, Sergeant Morales, kept his head down and his tempo steady; another, Private Harlan, surged early and faded fast.

They came back ragged and red-faced. Jack marked times, his pencil blunt and accurate. "Morales — good pace. Harlan — work on discipline, not fireworks. Recover, drink, and then we move to the fieldcraft lanes."

Fieldcraft began with camouflage and concealment. Davis and Kim watched the candidates disappear into a scrubby slope one by one, tasked with building a hide site large enough for two and camouflaged against a rock outcropping. They had twenty minutes with a limited kit: poncho, cord, a candle, and a length of burlap. No metal tools. The ground was hard, and the wind wanted to blow everything flat.

When Jack inspected the hides, he ran his gloved hand along seams and flaps, looking for loose edges, reflected surfaces, things that might catch a searchlight. Morales had buried his shelter low and used natural driftwood to break the outline; Cpl Ortiz had overbuilt, making a roomy hole that would be comfortable but would collapse under rain. "Comfort kills in the field," Jack said curtly. "We'll leave comfort to city boys."

Next came movement drills: bounding overwatch, silent wedges, crossing open ground without silhouette. The men moved in pairs and threes, Jack watching how they read terrain—where they hugged a fold, where they used a dip to mask movement, where they misjudged a slope and stamped their feet in the open. Walker timed their crosses and Davis called corrections over the training net. Radios clicked on and off like metronomes; discipline on the air mattered as much as discipline on the ground.

The decision drills were short, sharp, and unkind. Volunteers were presented with simulated contacts — a sudden flare, a mock machine-gun report, a screaming single round (courtesy of the Range Detail and ethically typed blanks). They had seconds to pick options: hold and observe, fix and withdraw, or press and capture. Jack watched who froze and who made crisp orders. "You can't lead if you're buying time thinking," he told them later. "You make a choice, you own it, and you move."

At noon, the field was a chart of dents and tracks. Walker and Jack huddled over the forms, comparing times and notes. "We need men who can shrink into the ground and do the arithmetic of patience," Walker said. "Not just brawn."

Jack nodded. "And radio discipline. Morales, Ortiz, Lee — they have the heads for it." He tapped a pencil against his lip. "Harlan's gutsy, but he runs hot. We'll see if he settles."

Selection wasn't an instant thing. They cut from twenty volunteers to twelve by the afternoon, then to the final two six-man rosters that would be molded into recon teams. Names were written, crossed out, rewritten; Walker argued with one call, Jack another. In the end the choices were pragmatic — men who moved without announcing themselves, who listened before they spoke, who fixed a snapped strap and kept pace.

By dusk the new teams were assembled in front of the Misfits. Each man looked smaller somehow standing shoulder to shoulder with Jack's veterans, but there was a steel to their jaws that had not been there at dawn.

"Congratulations," Jack said bluntly. "You made the list. Now the hard part starts." He pointed to the nearest rise. "PT begins at 0500 tomorrow. After that: night movement, hide-site construction under noise, radio suppression drills, immediate extraction rehearsals. We train tight so you can operate loose. You understand?"

They answered in the same clipped chorus— "Yes, Sergeant."

Jack watched his new men file to their assigned bunks. The Misfits stayed late, checking equipment, setting up a night lane, drilling comm procedures until the blue light of evening swallowed the ground.

"Push them hard," Walker said quietly as they packed up. "We've got three weeks to make ghosts of them."

Jack shouldered his empty pack and smiled a thin, tired smile. "They'll curse us before they thank us. That's the sign of good work."

Under the cold stars, Jack ran his hands through his gloves and felt the map of the ridge in his head like a promise and a threat. The furnace had been lit; the men who survived it would be shaped by it. Tomorrow will be harder. He welcomed it.

Three relentless weeks passed. Every dawn began with the same rasp of the whistle and the same barked order— "On your feet!" The hills echoed with the grind of boots, the rasp of shovels carving hide sites, and the muted click of safeties checked and rechecked.

Under Jack's watchful eye, the new recruits evolved from uncertain soldiers into disciplined shadows. They learned to move without sound, to read the terrain like a map written in instinct, and to vanish beneath snow, dirt, and darkness. Mistakes were punished not with anger, but with repetition—until every hand signal, every radio call, and every silent movement became second nature.

CPT Walker inspected them often but never interrupted. He let Jack mold them the way only a man who'd survived countless patrols could. The Misfits led drills, acting as both teachers and testers, pushing the recruits until exhaustion blurred into instinct.

By the final week, the two new teams were no longer recruits—they were recon men.

On the last morning of training, all three teams assembled in front of the command post. Their uniforms were caked in dust and sweat, faces smeared with streaks of mud and fatigue, but their eyes were sharp—focused.

Walker stepped forward, hands clasped behind his back.

"Three weeks ago, most of you were just line infantry," he began, his voice carrying over the wind. "Now you're part of Division G2 Recon—our eyes beyond the wire. What you see, what you bring back, will decide who lives and who doesn't. That's the weight of this job. Carry it well."

He turned toward Jack. "Sergeant Mercer, your teams are ready for tasking."

Jack nodded once. "Yes, sir."

Walker gestured to the two men standing beside Jack.

"Team *Lancer Seven*, led by SSG James."

James, a quiet, lean soldier with the look of a man who'd lived under fire, gave a curt nod. His team stood behind him, faces marked by confidence earned the hard way.

"Team *Lancer Eight*, led by SSG Horn," Walker continued.

Horn—a broad-shouldered veteran with a scar along his jaw—stepped forward. "Sir."

Walker met each of their eyes. "You'll be running recon operations under Sergeant Mercer's oversight. He'll coordinate all three teams. Lancer Six will handle forward leadership and mission planning. You'll report through me directly."

"Yes, sir," James and Horn answered in unison.

Walker gave a thin smile. "Good. Then as of now, all three Lancer teams are operational. Your first missions will come down within forty-eight hours. Rest, refit, and stay sharp. You're about to step into the next phase of this war."

As Walker dismissed them, Jack looked over the line of men—the Misfits, the newly minted Lancer teams, all standing shoulder to shoulder. Pride flickered in his chest.

They'd built something new out of nothing—three teams that could move like smoke and strike like thunder when needed. The hills would soon test them, but for now, the silence of the camp felt almost like victory.

Jack glanced toward Walker, who gave a subtle nod of approval. The job wasn't done—not by a long shot—but the foundation was solid.

They had built ghosts. And the ghosts were ready.

Epilogue

The wind over the Korean hills had grown quiet. The guns that once thundered day and night now slept beneath a sky stretched thin and pale, their echoes swallowed by distance and time. Snow still dusted the ridgelines, softening the scars of battle, but the land beneath it remained marked trenches filled, shell craters frozen like old wounds.

Jack Mercer stood at the edge of the ridge where the Misfits had first made their mark. Below, the valley spread out in muted gray, the same ground they'd crossed a dozen times in silence and fire. Now it looked peaceful—deceptively so.

He exhaled slowly, breath fogging in the cold air. Behind him, the camp buzzed faintly with the sounds of routine: distant hammering, the murmur of voices, engines growling somewhere down by the road. CPT Walker was still running G2 operations, the Lancer teams now fully integrated— three units strong, sharp, and capable. They'd earned respect from everyone who mattered, and more importantly, they'd survived.

Davis approached quietly, his ever-present notebook tucked under one arm. "Orders just came down, Sarge," he said. "We're standing down for forty-eight hours. Division says things are holding steady along the line."

Jack nodded. "That's a first."

"Yeah," Davis said, smirking faintly. "Walker said it feels wrong when it's quiet."

Jack didn't disagree. He glanced down again at the valley, then back toward the camp where the new Lancer teams were drilling. Horn's men moved like shadows; James' team was a mirror of the Misfits themselves. For a moment, Jack allowed himself to feel something close to pride.

"They'll do fine," Davis said quietly, following his gaze.

"Yeah," Jack replied. "They will."

Silence settled again, the kind that always came after too many losses and too many close calls. The kind that made a man remember.

Jack's thoughts drifted—through the smoke and snow, through the nights of whispered orders and the flash of muzzle fire, through the faces of men he'd led and buried. Then, beyond the war, to Patricia. Her smile in the garden. Her warmth against the cold. The promise they'd made before he boarded the C-47 back to this frozen place.

He reached into his pocket and felt the worn edges of her last letter. He didn't open it; he didn't need to. The words were memorized, every line a small light carried into the dark.

Walker's voice broke the quiet behind him. "Mercer."

Jack turned. The captain stood with his hands in his coat pockets, watching him.

"Division's impressed," Walker said. "Word is, once things stabilize, they might move G2 into Japan permanently. Could be you and your men finally get some real downtime."

Jack gave a faint smile. "We'll believe that when we see it, sir."

Walker chuckled. "Fair enough. But when that time comes—you've earned it."

Jack nodded, his eyes returning to the horizon. The war wasn't over. Maybe it never truly would be. But for the first time in months, he felt something shift—a sliver of calm, a moment of stillness he hadn't known since the parachutes of Normandy filled the night sky.

He turned back toward camp, the wind tugging at his coat, and started walking. The Misfits were waiting—his team, his brothers.

And somewhere across the sea, Patricia waited too.

The hills would keep their ghosts. Jack Mercer had made peace with that.
He was one of them now.

That night, as the camp settled into uneasy quiet, the field radio crackled softly in the corner of the command post. Davis, half asleep, reached for it—but stopped when he heard the tone. It wasn't Division. It wasn't even Army channels.

Jack stepped in, rubbing fatigue from his eyes. "What is it?"

Davis handed him the handset, frowning. "Message came through on an encrypted frequency—one we've never used."

Jack lifted it to his ear. The transmission was faint, distorted by static, but the words that cut through froze the air in his lungs:

"Lancer Six… this is Archangel. Priority message. Stand by for new orders. Target of interest—classified Omega. Coordinates to follow."

Jack lowered the handset slowly; eyes meeting Walker's across the dim tent.

The war, it seemed, wasn't done with them yet.